IMAGINING THE UNIMAGINABLE

THE POETICS OF EARLY MODERN ASTRONOMY

58

Internationale Forschungen zur Allgemeinen und Vergleichenden Literaturwissenschaft

IMAGINING THE UNIMAGINABLE

THE POETICS OF EARLY MODERN ASTRONOMY

Ladina Bezzola Lambert

Rodopi

Amsterdam – New York, NY 2002

This thesis was accepted as a doctoral dissertation by the Faculty of Arts of the University of Zurich in the summer semester 1999 on the recommendation of Professor Dr. Hans-Jost Frey.

Le papier sur lequel le présent ouvrage est imprimé remplit les prescriptions de "ISO 9706:1994, Information et documentation - Papier pour documents - Prescriptions pour la permanence".

The paper on which this book is printed meets the requirements of "ISO 9706:1994, Information and documentation - Paper for documents - Requirements for permanence".

ISBN: 90-420-1578-0

Printed in The Netherlands

to Claude

to Claude

Table of Contents

Illustrations

Acknowledgements

I would like to thank several people for helping me to get this study under way and bring it to completion. I am most indebted to my two advisors, Hans-Jost Frey and Timothy J. Reiss, for their generous support of my work and the stimulating discussions, which often had to be carried on across continents and an ocean. In particular, I am grateful to Hans-Jost Frey for his readiness to support a project that was outside of his own field and for following it with his usual extraordinary critical acuteness and curiosity; Timothy Reiss has overwhelmed me with his openness to take interest in the work of someone who came as a complete stranger. His incredible expertise in the fields of science studies and literature has been a rich and ever open source. The help of both has been invaluable. During a year I spent as a visiting researcher at Harvard University, Mario Biagioli was very influential in the early stages of research and writing. I thank him for many valuable comments and recommendations for critical reading. I am grateful to Allen Reddick and Jimena Canales for reading and commenting on parts of this study at different stages and to Andrew Shields for helping me with the final revisions. I also thank my brother-in-law René Lambert for the beautiful layout of the cover design. Finally, my thanks go to my family and, especially, to my husband, Claude Lambert, who saw me through it all and who has been a great companion. His unwavering dedication to teaching me some rudiments of time management has not been totally wasted.

I also thank the Swiss National Science Foundation for financing my year of research and studies as Harvard University.

Introduction

> Così, e molto più, può accadere che nella Luna, per tanto intervallo remota da noi e di materia per avventura molto diversa dalla Terra, sieno sustanze e si facciano operazioni non solamente lontane, ma del tutto fuori, d'ogni nostra immaginazione, come quelle che non abbiano similitudine alcuna con le nostre, e perciò del tutto inescogitabili, avvengachè quello che noi ci immaginiamo bisogna che sia o una delle cose già vedute, o composto di cose o di parti delle cose altra volta vedute; chè tali sono le sfingi, le sirene, le chimere, i centrauri, etc.[1]

> Thus, and more so, might it happen that in the Moon, separated from us by so much greater an interval and made of materials perhaps much different from those on Earth, substances exist and actions occur which are not merely remote from but completely beyond all our imaginings, lacking any resemblance to ours [i.e., our world] and therefore being entirely unthinkable. For that which we imagine must be either something already seen or a composite of things and parts of things seen at different times; such are sphinxes, sirens, chimeras, centaurs, etc.[2]

The famous passage from Galileo's *Dialogue* gives testimony to a claim I would like to raise here: the claim that Copernican cosmology lent new importance to the role of the imagination, even as it implied the moment of its crisis. Copernicus' reorganization of the cosmos, in which the Earth had come to occupy the position of a planet, implied the potential existence of other Earthlike worlds in a vast and, as later measurements and discoveries suggested, perhaps even boundless universe. It thereby transported the wider universe into the realm of the imaginable. At the same time, the suspicion arose that, in a universe that does not revolve around a central Earth, human imaginings might be too caught up in earthly dramas to grant access to other realities; that they were beyond human imagination.

As we all know, Copernican cosmology and the theories it generated concerning the plurality of worlds and the vast dimensions of the universe profoundly unsettled Western European thought and religion: It cast doubts on such articles of Christian faith as the doctrine of salvation and the certainty of divine governance over all earthly matters. It also raised questions about the nature of Creation and its relation to a transcendental truth, a divine Maker, questions that were of fundamental importance to

---

1 Galileo Galilei, *Opere*, ed. Antonio Favaro (Florence: Barbèra, 1890-1909), vol. 10, 289.

2 Galileo Galilei, *Dialogue Concerning the Two Chief World Systems*, trans. Stillman Drake (Berkeley: University of California Press, 1967), 61. Explanation mine. Throughout this study, I will give quotations from the principal primary texts I focus on both in the original language and in English translation. For reasons of space, I will do so only with quotations in the main body of the text. Quotations in the footnotes will be given only in English translation.

human identity. This loss of ontological, theological and, more generally, philosophical certainties "left a need for new understandings of divine, human and worldly relations".[3]

The human faculty of imagination was traditionally believed to involve just such relations between the divine, the human and worldly objects. And yet, the impact the new cosmology and its wider implications have had on the imagination – specifically the ability to visualize, to create visual images in the mind – and on its creative potential is a subject which has been little addressed.[4]

The finite geocentric cosmos endorsed by the Church was largely an amalgam of (Neo)Platonic philosophy, Aristotelian cosmology and Christian theology. Its most conspicuous characteristic was the distinction between a terrestrial realm opposed to a metaphysical-transcendental one in the heavens, a distinction one could basically characterize with such opposites as 'eternal' versus 'transitory', 'unchanging' versus 'changing', 'perfect' versus 'degenerate'. Though not in itself incompatible with materiality, the celestial realm, which contained the planets[5] and the fixed stars, was believed to consist of so superior and pure an essence as radically to defy any comparison with earthly matter. It was governed by a separate set of physical laws and shared in the nature of the divine: eternal, unchanging and perfect. As a physical world, the heavens were therefore just as inscrutable for human beings as the metaphysical-transcendental realm of God and divine truth. While emphasizing this radical *essential* difference (in the literal sense of a difference in essence), Christian theology maintained that the two realms were at the same time intimately related in their *spiritual* significance, since the whole cosmos was an expression of God's mind, with man created in God's image.

---

3 Timothy J. Reiss, *Knowledge, Discovery and Imagination in Early Modern Europe. The Rise of Aesthetic Rationalism* (Cambridge: Cambridge University Press, 1997), 79.

4 An exception is the pioneering work of Marjorie Hope Nicolson, especially her book *Science and Imagination* (Ithaca: Cornell University Press, 1962). Rather than dealing with the role of the imagination as such, however, Nicolson is here exclusively concerned with the way astronomical discoveries became an important source of inspiration for literary themes in the seventeenth, eighteenth and nineteenth centuries. A more recent notable exception is Fernand Hallyn's *The Poetic Structure of the World: Copernicus and Kepler* (New York: Zone Books, 1990), where he argues that Copernicus and Kepler established their cosmological hypotheses both on the structures supplied by rhetorical tropes (specifically metaphor, irony, anagogy and synecdoche) and on the basis of stylistic and thematic terms characteristic of specific literary genres. These are given a priori status.

I am using the term 'imagination' in its restricted meaning of 'visualization in the mind'. This corresponds to the traditional concept of imagination and the etymological significance of the term *imaginatio*, from which the modern translations 'imagination', 'immaginazione', 'Einbildungskraft', etc. were derived. In defining the term so narrowly, I do not mean to deny that the potential of the human faculty of imagination includes more than just sensual imagination (besides visual also auditory, olfactory, gustatory and tactile) and that there are functions of the imagination that are of a more deeply psychological and ontological nature; yet these are not what I am concerned with here. Moreover, my use of words as 'imaginative', 'imaginary', etc. may appear a little unorthodox at times, yet such terms should be understood in relation with my narrow definition of 'imagination' as 'visualization'.

5 Beside Jupiter, Saturn, Mars, Venus and Mercury, the Sun and the Moon were also counted among the planets.

According to Christian belief, the originally Platonic distinction between the terrestrial and the metaphysical world was reinforced by human sin. With man's fall from grace and the resulting degeneration of man and nature, the relation between terrestrial nature and divine truth ceased to be obvious and remained only arcanely manifest. Hence, during their life on Earth, human beings could gain knowledge of the true nature of the cosmos only to the extent their flawed faculties permitted. Full insight into God's mind, the metaphysical realm as well as the physical universe, would eventually be theirs in the hereafter. What they perceived through their senses related to a secret truth, but the relation that was left was complex and exclusively spiritual in nature. Thus, they could not take their sensual perceptions at face value. There was little use in attempting to describe the universe in its physical essence: In its degenerate earthly form, it either remained unknowable or it was misleading, so that only its symbolic meaning could give the human mind an intimation of its significance to divine creation.

Rather than the *physical* condition of matter in the universe, it was the *spiritual* significance of things and their relation to the Creator that merited scrutiny.[6] In its spiritual significance the cosmos was considered one in all of its parts. All levels of existence were intimately connected through an elaborate hierarchical system of analogies and correspondences, extending "from the Mushrome to the Angels".[7] Each part of the cosmos thus implied the whole, though in different degrees of perfection. It was in the attempt to trace this obscure relation that the imagination became useful.

The Christian concept of imagination, as it had been gradually developed up to the sixteenth century, was inextricably connected to this view of the cosmos. Its three basic points of relation to this cosmology are:

(1) the fundamental oneness and finitude of the cosmos;

(2) a conception of alterity present in the distinction between Earth and heaven which combined a notion of arcane relation with a notion of complete otherness;

(3) the point that in the Christian cosmos the heavens with its planets and stars were considered not so much a physical as a spiritual place.

The theoretical foundation of the Christian concept of imagination has its origins in the theories of Plato and Aristotle. Plato's distinction between the transitory material world of becoming and the world of being, i.e., the metaphysical realm of transcendental Ideas, directly relates to his conception of man and the distinction between body and

6 We find this also reflected in the fact that the pre-Copernican cosmological system was a system of appearances. It did not claim to present the cosmos as it is in reality, but as it appears to the human eye. Claims regarding the true nature of creation could only be based on the Word of God, and this was the domain of the theologians. See chapter 2, pp. 32-33 and notes for bibliographical references.

7 Samuel Ward, *The Life of Faith*, 3rd ed. (1622), quoted by C.A. Patrides in his introduction to *Sir Thomas Browne. The Major Works* (Harmondsworth: Penguin, 1977), 26.

mind. Whereas the body and its senses are bound to the material world, the mind partakes of the divine and has access to transcendental truth.[8]

In this dualistic system, there is very little room for the imagination. What Plato terms *phantasia* is a combination of sense perception and the evaluative judgment according to which perception figures in the mind.[9] *Phantasia* is thus responsible for the picture we form in the mind of what we see while we see it. Though Plato distinguishes *phantasia* from both our mental faculty and from perception, its link to sense perception is immediate. He grants the imagination little creative value, condemning it as a pseudo-world of imitations formed out of mere copies of the copies that constitute the material world. What human beings imagine therefore resides at a third remove from truth and can never have access to truth. Reason, on the other hand, does not depend on images to attain knowledge; its relation to knowledge is one of innate affinity.[10]

The importance of Plato's theoretical account of imagination for later interpretations that would incorporate the concept into a Christian framework lies primarily in his integrating the notion of imagination as *mimetic* representation into a cosmos defined by *mimesis*. The image was, by definition, a reproduction pointing to an original beyond itself – as was the world. Plato thus created the stage on which imagination was to perform a more active, if no less constricted, role in a Christian cosmos.

The other fundamental source of influence for the theory of imagination was Aristotle. Shifting the terrain of investigation from *metaphysics* to *psychology*, Aristotle concentrates on the properly psychological status of the image as a *mental* representation.[11] In contrast to Plato, Aristotle defines *phantasia* as an internal activity of the mind which mediates between sensation and reason. "The image serves as the internal representation of sensation to reason."[12] Since, according to Aristotle, ideas in the mind do not derive from a super-sensible realm (as Plato maintained), but by

8 *Timaeus*, in *Plato in Twelve Volumes*, trans. Rev. R. G. Bury (Cambridge: Harvard University Press, 1965) , vol. 9, 181-91. The same distinction is also made in the *Republic*.

9 J. M. Cocking, *Imagination: A Study in the History of Ideas* (London: Routledge, 1991), 12-13.

10 Richard Kearney, *The Wake of Imagination: Ideas of Creativity in Western Culture* (London: Hutchinson, 1988), 87-99. I am simplifying Plato's position for the sake of a clearer argument. As Kearney points out, "Plato's metaphysical assessment of imagination is not ... entirely negative." There are passages in the dialogues that are inconsistent with his general evaluation; thus in book VI of the *Republic* he "concedes that knowledge (epistêmê) may at times have recourse to what he terms 'thought-images' in order to enable our human understanding (dianoia) to give figurative expression to its abstract ideas. Such images may serve a positive heuristic purpose by providing the mind with quasi-intuitive representations of otherwise invisible truths; they help us to visualise and thereby communicate the concepts of knowledge." (99) Here Plato only touches upon a psychological function of imagination as internal modes of thought which was to be given a comprehensive treatment by Aristotle in *De anima*. (102) See also Reiss, forthcoming, where Plato's more critical understanding of *mimesis* in the *Republic* is contrasted with that presented in the *Phaedrus*, which Reiss (together with critics like Melberg, Gebauer and Wulf as well as Prendergast) terms "a creative play ultimately allowing the soul at least some access, be it only by imagination, to [the substantial] reality [of Ideas]." Cf. *Against Autonomy: Global Dialectics of Cultural Exchange* (Stanford: Stanford University Press, 2002), 154. Explanation mine.

11 Kearney, *Wake*, 106.

12 *Ibid.*, 107.

abstraction from the sensible world itself, images become the indispensable mediator between perception and reason.[13] Even though imagination still belongs to a part of the mind inferior to reason, it is now part of our mental faculty and "its mediational rapport with our sensible experience is one which may result in the apprehension of truth rather than leading us into an illusory world of imitations."[14]

Moreover, Aristotle defines a wide spectrum of experiences in which *phantasia* occurs and where it stands in a mediate rather than an immediate relation to perception. They include memory, recollections, dreams, hallucination and – most fundamentally – thinking.[15] In all of these mental activities, *phantasia* draws upon memory – conceived as the storehouse of images (*phantasmata*) – to evoke past perceptions. Aristotle here alludes to a creative potential of imagination that seems closer to our modern understanding of the term. Yet, as Alan White points out, there is an important difference between Aristotelian interest in imagination and the contemporary one:

> though he does mention our power to call up a *phantasma* whenever we wish, and also to combine several *phantasmata* into one, he lays little stress on its use to envisage possibilities other than the actual. More prominence is given to the passive reception of appearances than to the active power to call them up.[16]

Aristotle as well as his medieval and Renaissance interpreters still neglected the notion of imagination as a fiction-making power. The idea of imagination as a creative power that voluntarily evokes sensual perceptions of the past and freely separates and recombines these images into new fictions would reappear only much later with philosophers of the early modern period such as Hobbes, Locke and then especially Hume and Kant. This is not a coincidence. For interpreters of the concept of imagination within a Christian framework, the notion of imagination as a creative power was of little value. Their interest lay in the question of whether and how the world as they perceived it related to a higher truth rather than how images could be turned into other images. Such a practice would lead only into idolatry.

Of far greater significance to them was Aristotle's argument that *phantasia* serves as a mediator between sense perception and reason, and that the mind can have access to truth only through images. Christian philosophers integrated this argument into a dualistic cosmological system of Platonic origin, which posits a transcendental realm of divine truth. Ironically so, for this meant that Aristotle's theory was detached from its psychological context and reapplied to metaphysics.

Christian theology for the most part adopted Plato's distrust of imagination as a creative faculty. At the same time, "the Christian doctrines of Incarnation and of redemption encourage[d] the view that creation, seen with the illumination of grace, is

---

13 Cocking, *Imagination*, 18. Cocking explains the difference from Plato as follows: "In Plato *phantasia* is simply a word to refer to our awareness of the sensible world when our mind has taken it in, in Aristotle *phantasia* becomes the handmaid of *nous* – an elusive and intriguing power of the mind mediating between sensation and thought." (20)

14 Kearney, *Wake,* 109.

15 White, *The Language of Imagination* (London: Blackwell, 1990), 7-9.

16 *Ibid.*, 9.

itself a glorious manifestation of the divine."[17] God had created the world in His image, and it was therefore in the image of worldly things that a relation to God and to truth manifested itself, however obscurely. While man certainly needed to be prevented from idolatrously indulging in vain images, his highest vocation was to 'mirror' the Supreme Artist of the universe faithfully. This was the task of the imagination.

The concept is best explained with reference to an influential Neoplatonist theory of creation, the theology of emanation, which J. M. Cocking considers to be a vital step in the metaphysical promotion of *phantasia*. The theory explains creation "through successive emanation of the divine substance – God, or the One, radiating outwards in waves reaching further and further away from the original perfection and creating the hierarchy of substances from pure spirit to base matter." We noted earlier that the psychological tradition springing from Aristotle regarded *phantasia* as the intermediary between sensation and reason. "[I]n the theology of emanation it becomes the intermediary in the reverse process; the divine mind, creating, changes from pure Intelligence through imagination into sensible substance."[18] To gain insight into the mystery of Creation, the human creature needs to reverse this process once again. As the *simulacrum* of the Divine Creator, he needs to reenact creation by construing the created world as a set of mirrors, all of which point towards God.[19] In doing so, the creature must disregard the material substance of things as being incompatible with the more sublime parts of creation, and reflect exclusively on their image and the transcendent spiritual significance that manifests itself through it. The images of things carry symbolic value; they are the crude representations of a complex, inexpressible truth, a truth that human beings can only divine with the help of reason and grace. Imagination then becomes the faculty that combines sensations into the total pattern of our perceptual world and, with the help of reason, reconstructs the intelligent pattern to which the world owes its origins.

According to the Neoplatonist Marsilio Ficino, imagination is part of a process that serves to redeem the fallen world in which the Platonic forms have become contaminated with matter:

> the Soul of man, which is affected by the individual impulses of individual bodies through its earthly body, receives these images of the Ideas maculated by the matter of the universe through perception, but collects them through fantasy, cleans and refines them through reason, and connects them at last with the universal Ideas of the mind. So the celestial ray that had descended to the lowest things returns to the higher beings, because the images of Ideas, formerly dispersed in matter, are collected in the fantasy, and, formerly impure, they are purified in reason, and, formerly particular, they are lifted in the mind to universality. In this way the Soul of man restores the world that had already been shaken. For by its gift the world that was formerly spiritual and had become corporeal is continuously being purified and daily becomes spiritual.[20]

---

17 Cocking, *Imagination*, 69.

18 *Ibid.*, 52-53.

19 The concept is taken from St. Bonaventura. (Kearney, *Wake*, 124)

20 Quoted in Cocking, *Imagination*, 175.

In Ficino's account, the function of imagination is clearly subservient to that of reason. In contrast, and with the risk of complicating the overly neat lines of this history a little, we must turn to another famous practitioner of hermetic art and the much more powerful role he attributes to imagination in the mind of the Magus. Giordano Bruno makes a radical claim for the primacy of imagination among all faculties in the cognitive process. With the help of the art of memory, imagination becomes "man's highest power, by means of which he can grasp the intelligible world beyond appearances through laying hold of significant images", stored in his memory.[21] Bruno thus establishes imagination – rather than reason – as the means through which the divine microcosm (man) can reflect the divine macrocosm (the universe), can grasp its meaning from above and, indeed, control it.[22]

The centrality of imagination in Bruno's magical art, on the one hand, and, on the other hand, Bruno's role of a fervent supporter of Copernican theory and the theory of an infinity of worlds to some extent blur the distinction between a pre-Copernican and a post-Copernican role of imagination which I am trying to outline here. Bruno is a Janus-faced figure, for his involvement in Hermetic magic also relates him back to a pre-Copernican dualistic cosmology in which man can hope to gain only *spiritual* access to celestial powers. However, the rise in the hierarchy of mental faculties which imagination experiences in Bruno's system in relation to reason changes little about the fact that the task of the human mind still lies in unifying the world by mirroring God's creation. When he attributes the creative task of establishing the links in the cosmic pattern largely to imagination rather than reason, Bruno merely enhances the role of imagination in the old process.

Man's imaginative creations were not original acts of creation, but mere imitations of God's original creation of the cosmos. Bruno's description of mental mirroring is more radical than the traditional ones in as much as in his theory the imitation acquires more clearly competitive traits. In its general framework, it remains faithful to the traditional paradigm. As Richard Kearney comments in his genealogical history of imagination:

> in Western classical and medieval culture ... we find that the imagination is never understood in terms of some internal, subjective power of man alone. It is always evaluated in terms of man's relation to a divine or cosmic power greater than imagination itself.[23]

In such a closed cosmological and philosophical system, human beings used the imagination to bring the world into its own and preserve it in its oneness rather than as a tool to visualize alternatives to the world as they knew it. To do so, the spiritual links between the different realms had to be made evident through images of analogy in which terrestrial phenomena came to represent the transcendental realm of truth symbolically. Accordingly, the preeminent mode in medieval literature and scriptural exegesis is allegory, where the imagery of the fable is composed of elements from the

---

[21] Frances Yates, *The Art of Memory* (London: Routledge and Kegan, 1966), 230. See also *idem*, *Giordano Bruno and the Hermetic Tradition* (London: University of Chicago Press, 1964).

[22] Yates, *Memory*, 158.

[23] *Wake*, 83.

material and familiar world, to which the reader may attribute a spiritual, moral or psychological signification through interpretation. By focusing on the alleged symbolic significance of material objects, allegory demonstrates how everything in the cosmos analogically relates to everything else. Place is thus defined not so much in physical as in spiritual, psychological or moral terms.

This changed quite radically with the embracing of Copernican theory. The new theory deprived the Earth of its central position in the universe, turning it into one of the planets that orbit the Sun. It thus opened up the possibility of a material similarity between the Earth and the other planets. The heavens were now no longer conceived in terms of a fundamental alterity, but became accessible to the human mind as a physical world like the Earth. At the same time, the Earth could no longer serve as the point of reference (however flawed) for the entire cosmos. With the world bursting open, the search for knowledge was directed outward and turned into a quest for the discovery of new worlds. The essential difference posited between the terrestrial and the celestial realms disappeared. It was replaced not only by the concept of an essential similarity linking the Earth to the rest of the universe, but by a new potential for difference that emerged with the possibility of alternative worlds.

An important factor in the scenario of this conceptual change was Galileo's introduction of the telescope to astronomy in 1609. By revealing a multitude of previously unperceived phenomena in the sky, this instrument furnished evidence of a universe of much vaster dimensions than hitherto suspected, but also – and more fundamentally – of a universe that stretched beyond the reach of human vision. Accordingly, it deprived the cosmos of a unity that had been of basic significance to the Christian-Aristotelian view of the world. In addition, the telescope also permitted closer scrutiny of 'old' phenomena. The Moon, with its dominant position in our night sky, was the principal target for exploration with the new instrument. It was unique among the bodies in the heavens in that its image through the telescope revealed characteristics which made it possible to describe the Moon by analogy to the Earth. It thus offered support for the principle of a material similarity between the Earth and the rest of the universe. Another, even more revolutionary discovery was the moons of Jupiter. In the company of its moons, Jupiter offered a model of the Copernican system in miniature. Such a system moreover disproved the uniqueness of the Earth with its orbiting Moon, making the Earth planet-like. The discovery of Jupiter's moons thus offered a strong argument in favor of both Copernican theory[24] and the existence of other worlds, the latter argument suggesting itself especially in combination with the evidence of the Earth-like nature of the Moon.

Yet while showing more than the naked eye, the telescope also left much implied. It did not offer a detailed picture of the Moon, much less, of course, of the other planets and stars. In fact, Galileo's drawings of 'how he saw' the Moon bear little resemblance to any modern mapping of the Moon's surface. Nor did the telescope affect the physical inaccessibility of these objects; indeed, in some ways it rendered the study of the heavens more problematic in that it constantly implied objects and spaces beyond its

[24] For a discussion of how the discovery of Jupiter's satellites was made instrumental in the defense of Copernican theory, see Timothy Reiss, "Espaces de la pensée discursive: le cas Galilée et la science classique," *Revue de Synthèse* 85-86 (Jan. – June 1977): 6-7.

reach. These blanks had to be filled with the help of analogical reasoning and the imagination.

The combination of the fact that the telescope revealed less than it implied with the principle of a material similarity between the Earth and the Moon (and, by implication, other heavenly bodies) defined and justified a need to visualize these remote objects as physical places in the imagination. In consequence, both the significance and the role of the imagination underwent a fundamental change. Hitherto, the imagination had served as a handmaid[25] that helped the mind reconstruct the intellectual pattern of the cosmos by representing the stored visual impressions of the material world as symbols of an immaterial higher truth. Now, however, it could be used as a creative faculty which, by calling upon, separating and recombining past impressions, actively participates in the creation of alternative material worlds. Here the imagination offers both potential and liability: the projection of familiar images onto the unknown can render the strange conceivable, but in the visualizing of the unknown the imagination is rigidly confined to past perceptions. Even though the imagination may modify actual perceptions by separating and recombining them, it nevertheless cannot create anything fundamentally new, but only recycle old material. As a result, the creations of the imagination are essentially characterized by metaphor.

As long as the focus is on the similarity between the Earth and other heavenly bodies, this is not a problem; yet when it comes to conceiving alternative worlds and imagining ways in which they might differ from the familiar world, the dependence on actual experiences proves inhibiting. The mere possibility of the worlds out there being quite different from the world we know renders the creative potential of the imagination doubtful. The change in cosmology that originated from Copernican theory thus not only redefined the role and significance of the imagination as a concept, it also implied the moment of its crisis. One of the major concerns of this book, besides analyzing the creation of imaginary worlds, will be to see in what ways the crisis of the imagination as a creative potential manifests itself, and how it is confronted.

It is important to emphasize, however, that this change in the significance and role of the imagination applies only to the context of the cosmology of Platonic, Aristotelian and Christian heritage. Alongside the dominant Platonic conception of mimesis and a mimetic imagination, Timothy Reiss has traced in Western antiquity another conception of mimesis and its function in art, one which attributed to the imagination a much more creative role. Reiss recalls a passage in one of Seneca's *Epistles*, where he adopted Plato's image of the artist as a bee, which, culling nectar from various flowers, makes honey from it. "'It is not certain', [Seneca] went on, 'whether the juice which they obtain from the flowers forms at once into honey, or whether they change what they have gathered into this flavor through some mixture and property of their breath [spiritus]'."[26]

According to Reiss, later "normative interpretation" reconciled this passage with "dominant forms and categories", thus effecting a "cancellation ... of [Seneca's] more

---

25 In a witty allegory, Richard of St Victor likens contemplation to a mistress who inhabits the inner sanctuary of reason, and imagination to a handmaid who frequents the outer rooms of bodily desire. (Kearney, *Wake*, 120)

26 *Against Autonomy*, forthcoming, 161. Comment T.R.

complex notion of mimesis". Seneca used the image of the artist as a bee "to show how the artist absorbed elements from many different 'flowers' to create a quite different amalgam and compound. ... His bees were makers, fictive imaginers in the most essential ways: they created new forms and experiences." Though the artist's creative act involves copying, in the sense of "following, imitating, emulating – previous writers" (i.e., what the normative interpretation had taken to be the exclusive meaning of the passage), and also copying from nature – what makes his or her work creative is the way it turns "potentially 'discordant voices' ... into a single harmonious soul". This denotes "human creativity in all of its forms and at its most originary level."[27]

I am quoting Reiss' argument at such length here because his interpretation of Seneca's bee image includes both the notion that the building materials of the imagination are essentially given as well as the notion that old fragments are incorporated into something new. The Senecan image of the artist's creative act thus also binds the artistic imagination to the familiar world, yet within its limited territory the imagination still finds ways to be innovative. The central chapters of this book offer readings of seventeenth-century astronomical texts that focus on the dependence of imaginative creativity on experience (both as phenomenon and theme) and strategies devised to deal with it. However, the concept of mimesis with innovation as it is suggested in Seneca's bee image will reappear in the first and last chapter of this study, where I will discuss the limitations and creative possibilities of the imagination in a literary context. Before introducing the contents of this study, I am briefly going to turn to the question of source texts and intertextual relations.

The attempts to imagine alternative realities beyond the familiar terrestrial one provoked by the Copernican revolution and, more particularly, by the invention of the telescope, had certain classical prototypes that became important models for seventeenth-century cosmic fantasies. Four texts were particularly influential in this respect: Cicero's *Somnium Scipionis*, Plutarch's *De facie in orbe lunae*, Lucian's *Ikaromenippos* and the *True History*.

Cicero's *Somnium Scipionis* offers a cosmic view of the Aristotelian geocentric cosmology of solid spheres that emphasizes the comparative insignificance of the Earth when it is considered in the context of the whole universe.[28] Although to a seventeenth-century reader Cicero's text merely illustrated the by-then canonized cosmology, the cosmic perspective of the universe that it offered became an important source of inspiration to early modern authors. Kepler, moreover, adopted the theme of the dream vision in his own *Somnium*, a point that I will be returning to in the third chapter.

Certainly the most influential classical text was Plutarch's *De facie in orbe lunae*, where, in a philosophical dialogue, one of the interlocutors (arguably Plutarch's fictive persona) proposes the thesis that the Moon's surface is rough and full of mountains and valleys and presents the Moon as a counterworld to the Earth. Plutarch became a central authority for defenders of Copernicanism, who used the argument of relativism against the geocentric worldview.

---

27 *Ibid.*, 161-65. Explanation mine.

28 Marjorie Hope Nicolson, *Voyages to the Moon* (New York: Macmillan, 1948), 17.

Lucian's *Ikaromenippos* and the *True History* give fantastical accounts of the world in the Moon and its inhabitants. The latter even includes cosmic warfare, in which Endymion, king of the Moon, commands the lunar troops. Although Lucian's playful fantasies could obviously not serve as a serious source of inspiration for astronomers and cosmologists,[29] they nevertheless offered fictions of alternative physical worlds that enact the same kind of imaginative play of combinations found in several seventeenth-century astronomical fictions. Kepler, for example, adopted the idea of the enormous size of lunar creatures from the *True History*. More importantly, however, Lucian's texts became a rich source for cosmic fantasies of a more openly fantastical kind.

My concept of imaginative creativity and its limitations requires at least some passing reference to the theories concerning the creativity of the imagination formulated by such writers as Fichte and Schelling and adopted by some early Romantics, particularly Schlegel, Novalis, Coleridge and Wordsworth. In referring to these writers, I simply wish to draw a distinction between my own notion of the creativity of the imagination and the one developed in these theories. According to the theory of imagination that Schelling develops in his *System of Transcendental Idealism*, the creative imagination is the faculty by which we may intuit a deeper spiritual truth beyond the outward appearances of objects in the world. Abolishing the sharp distinction Kant had drawn in the *Critique of Pure Reason* between the appearances of things and their true identity, which is forever beyond human understanding, Schelling and other idealists argued that the appearances of things were symbolic of the deeper truth of things, i.e., that they participated in this truth.[30]

In contrast, the following discussion works with a much simpler notion of imaginative creativity – one based not on the ability to transcend appearances and conceive the ultimate nature of the world, but rather on the ability to create *alternative appearances*, that is, fictions of alternative material worlds, and the inevitable restriction of such fictive creations to past perception. If the imaginative descriptions of other worlds in the texts I will be looking at are presented as metaphors, then they stand for potentially different *appearances*. These appearances are unimaginable, not because they are incompatible with a sensuous image, but because the creations of the imagination are bound to the familiar. The imaginative descriptions are metaphors for alternative appearances, not symbols of a deeper truth. In the imaginative creation of alternative worlds, the restriction of the imagination to past experience is a 'metaphysical' problem only in so far as the imagination cannot reach beyond the sensuous impressions that are available to it.

Much more relevant to my argument is the definition of human wit offered in some seventeenth-century treatises on rhetoric. Of particular interest for our purposes is Emanuele Tesauro's monumental work *Il cannocchiale aristotelico*, for a long time the

---

29 In the Preface to the *True History*, the reader is explicitly warned not to believe a word of what is said. Thus, the narrator presents himself as an "honest liar" and declares: "I am writing about things entirely outside my own experiences or anyone else's, things that have no reality whatever and never could have. So mind you do not believe a word I say." Cf. Lucian, *Satirical Sketches*, ed. and trans. Paul Turner (Bloomington: Indiana University Press, 1990), 250.

30 Mary Warnock, *Imagination* (London: Faber & Faber, 1976).

most influential study of Baroque poetics. The strange conceit of the book's title – the 'Aristotelian telescope' – refers initially to Aristotle's *Rhetoric*, which for its great analytical power "may be called a most clear TELESCOPE to be used to examine all the perfections and imperfections of Eloquence."[31] At the same time, the metaphor of the telescope also refers to the power of wit (*ingegno*), which Tesauro declares to be the distinctive characteristic of all superior literary texts.

Ingegno "is to be divided into two aspects: *perspicacia* and *versabilità*. 'Perspicacity' ... 'penetrates the farthest and most minute circumstances of every subject.' At the same time, 'versatility [but also 'reversibility'] speedily compares all these circumstances among themselves and with the subject: it connects and separates them; increases or diminishes them; derives one from the other; outlines the one with the other or with marvellous dexterity puts one in place of the other as the juggler does his stones.'" The result is "clever conceits and subtle metaphors: which is really all forms of metaphor in general."[32] Wit works like a telescope in that it may be used to scrutinize one's surroundings in the attempt to establish as many different qualities in each object as possible. By comparing these qualities and establishing similarities between them, the ingenious mind creates ever new metaphors.[33]

The terms of comparison are not essential qualities of things, but merely their attributes. Accordingly, for Tesauro (and other seventeenth-century theorists of wit such as Gracián, Matteo Pellegrini and Sforza Pallavicino), "all forms of wit arise from the artificial likenesses or unlikenesses between properties in different categories of being."[34] Metaphor establishes similitude between dissimilar things; in fact, the greater the dissimilarity between the objects compared with each other, the wittier the metaphor. Tesauro calls such metaphorical conceits "argomenti urbanamente fallaci" ('urbanely deceitful arguments').[35] In his view, the acuteness of wit "is specifically not discovery or rediscovery but artifice."[36] Or in Timothy Reiss' terms: "Tesauro emphasizes that what is called 'truth' is therefore in its essence a 'lie': for what is 'known' of any object is but the consequence of these activities of mind."[37] The 'Aristotelian telescope' of the title – an anachronism that in the philosophical (and political) context of the seventeenth century amounted to an oxymoron – is a prime example of such an artificial construct. For Tesauro, all representation is potentially metaphorical, which is

---

31 "Talche possiam chiamar le sue Rettoriche Un limpidissimo CANNOCCHIALE; per esaminar tutte le perfettioni, & le imperfettioni della Eloquenza." Cf. *Il cannocchiale aristotelico*, ed. August Buck (Bad Homburg: Gehlen, 1968), 2-3. Translation mine.

32 Timothy Reiss, *The Discourse of Modernism* (Ithaca: Cornell University Press, 1982), 26-27. Comment T.R.

33 Tesauro's definition of *ingegno* corresponds to Aristotle's discussion of the creation of metaphors at the end of his *Poetics*: "But the greatest thing by far is to be a master of metaphor. It is the one thing that cannot be learnt from others; and it is also a sign of genius, since a good metaphor implies an intuitive perception of the similarity in dissimilars." Cf. *The Rhetoric and Poetics of Aristotle*, ed. Friedrich Solmsen (New York: Random House, 1954), 1459a 5-8.

34 A. J. Smith, *Metaphysical Wit* (Cambridge: Cambridge University Press, 1991), 51.

35 *Cannocchiale*, 489.

36 *Metaphysical Wit*, 51.

37 *Discourse*, 27.

also to say that all representation is indirect and thus, at least to a certain extent, insincere.

Eileen Reeves offers an interesting discussion of the book's frontispiece, showing that it constitutes an unmistakable reference to Galileo's *Letters on the Sunspots*. Indeed, she argues that Tesauro's *Cannocchiale* represents the "literary counterpart" to Galileo's *Letters* since the taxonomy that Tesauro follows in his analysis of metaphor corresponds to the categories Galileo uses in the *Letters* to examine the bright spots visible on the solar surface. In Galileo's argument, the spots function as metaphors to describe and define the "central enigma", "the real, unique, non-metaphorical sun, which may be figured forth metaphorically, or not at all."[38] Reeves relates this metaphorical representation of the Sun to Galileo's belief – set down in his third letter on sunspots – that the true substances of things are unknowable and that, consequently, all human knowledge necessarily depends on appearances. If human knowledge is limited to the appearances of things, this means that all representation is metaphorical. In what follows, the metaphorical nature of representation and its inability to represent 'truly' will be one of the central themes. Let me now introduce the contents of this study.

In the second and third chapter I am going to analyze the earliest descriptions of the Moon based on telescopic observation – Galileo Galilei's *Sidereus nuncius* and Johannes Kepler's *Somnium* – to show how they create the lunar world in the imagination by projecting familiar images onto a strange world. I will also use passages from the *Dialogo sopra i due massimi sistemi del mondo*, where Galileo offers an elaboration of his lunar theories. An analysis of the imagery of these texts will reveal the odd provenance of some of the images used. They refer to traditional ideas, ancient myths and fables that are quite incompatible with the astronomical argument presented. They appear in these texts solely because of their authority and emblematic power.

Alongside the images themselves, I am going to look at rhetorical strategies employed in the descriptions of the Moon that play an important role in the creation of the lunar world. The texts by Galileo and Kepler will first of all concern me because of the way they create the lunar world by projecting familiar images onto strange visions. However, they are also interesting for the way in which they manifest an awareness of the limitations of the imagination and attempt to do justice to the possibility of otherness by taking critical distance from the imagery they employ. I will try to show how – in quite different ways – both Galileo's lunar narratives and Kepler's *Somnium* suggest ways in which the images that are used to describe the lunar world, though limited in their particularity, are acceptable tools of representation as long as we read them metaphorically rather than literally.

The rhetorical construction of these astronomical texts and of the worlds described in them is one of my main concerns throughout this study. Here, my analyses of texts by early disciples of Copernican theory will confirm what Alan Gross has rightly termed Paul Feyerabend's "deepest and truest insight": that "early in the development of a new scientific world view, at a time when a new theory is seriously underdetermined by the evidence, 'style, elegance of expression, simplicity of presentation, ten-

---

[38] "The Rhetoric of Optics: Perspectives on Galileo and Tesauro," *Stanford Italian Review* 7 (1987): 129 and 138.

sion of plot and narrative, and seductiveness of content become important features of our knowledge.'"[39] At the same time, I am also interested in how the texts in question offer a critique of their own rhetoric, how they comment on their techniques of representation.

My use of the term 'rhetoric' may require a clearer definition: To the best of my knowledge, most studies focusing on the rhetoric of scientific writing have used the term according to a strictly classical definition, limiting it to its purpose of persuading an audience.[40] While acknowledging the classical *persuasive* function of rhetoric, I prefer to use the term in the broader sense that has been ascribed to it in modern theories of discourse and which addresses the *creative* function of rhetoric. In this sense, a rhetorical analysis of early astronomical texts focuses on the way a world is created both as text and created from texts, through intertextual links that bind it to tradition.[41] The idea of rhetoric creating a world discursively seems to me particularly apposite in the texts of early astronomy, which literally figure on a blank page. The critical perspective on their use of rhetoric which the texts examined here imply in

---

39 *The Rhetoric of Science* (Cambridge: Harvard University Press, 1990), 97.

40 See, for example, Alan Gross, *ibid.*, and Jean Dietz Moss, *Novelties in the Heavens: Rhetoric and Science in the Copernican Controversy* (Chicago: University of Chicago Press, 1993), as well as Maurice Finnochiaro, *Galileo and the Art of Reasoning: Rhetorical Foundations of Logic and Scientific Method* (Dordrecht: Reidel, 1980). See also Brian Vickers' article "Epideictic Rhetoric in Galileo's *Dialogo*," in *Annali dell'Istituto e Museo di Storia della Scienza di Firenze* 8, no. 2 (1983): 70-102, where he discusses Galileo's use of techniques of praise and blame, the establishment of the speakers' *ethos* as well as the use of different *personae*. In *Descartes and the Resilience of Rhetoric* (Carbondale: Southern Illinois University Press, 1990), Thomas M. Carr discusses the use of rhetorical devices made by Descartes and a series of thinkers who drew heavily on him. Though his emphasis is also on persuasion, Carr examines the theory of rhetoric of these writers rather than their performance or the tension between the two. Against Descartes' famous repudiation of rhetoric and his emphasis on self-evidence, he shows that Descartes' method leaves room for rhetoric if it is considered not in terms of its outward form but in terms of the notion of *attention*.

An important exception is James Bono's *The Word of God and the Languages of Men: Interpreting Nature in Modern Science and Medicine*. Vol. 1: *Ficino to Descartes* (Madison: University of Wisconsin Press, 1995). Bono rejects the notion of a sharp division between premodern European "bookish culture" and modern "scientific culture", arguing that the new science retained the central metaphor of the West: the trope of the Book (the Bible and the Book of Nature) and the mythical notion of God's authorizing word. In consequence, scientific culture changed only gradually as it adapted formative "master narratives", specifically biblical narratives that concerned themselves with the origins and nature of language. Another important exception is the work of Eileen Reeves, "Rhetoric", and her more recent book *Painting the Heavens: Art and Science in the Age of Galileo* (Princeton: Princeton University Press, 1997). However, Reeves focuses on the rhetoric of visual representation rather than on textual rhetoric as I do here.

41 Maurice Slawinski argues that beside serving as an important tool of persuasion, "rhetorical modes of thought contributed to shaping the very logic of scientific enquiry [in seventeenth-century scientific discourse]." "There is no simple dividing line between rhetoric as a system of persuasion and rhetoric as a comprehensive system of thought. To adopt rhetorical forms of argument so as to gain public credence invariably involves a rhetoricization of scientific logic itself." Cf. "Rhetoric and Science/Rhetoric of Science/Rhetoric as Science," in Stephen Pumfrey, Paolo Rossi and Maurice Slawinski, eds., *Science, Culture and Popular Belief in Renaissance Europe* (Manchester: Manchester University Press, 1991), 72 and 90. My addition.

their own structure suggests that the discursive worlds these texts create are offered as a world substituting a reality that remains unknown in its true identity. Rather than offering a direct representation of this reality, the world of the text serves as a kind of stand-in that circumscribes this reality and implies it as a potential.[42]

The fourth chapter of this book focuses on a text where intertextuality is not only the fundamental creative principle in the description of other worlds, but where this intertextual practice becomes one of the principal narrative themes. The text in question is Cyrano de Bergerac's *Estats et empires de la lune* and the *Estats et empires du soleil.* Cyrano's lunar and solar worlds, presented in the form of a novel in two parts, 'materialize' from a literary tradition and common figures of speech. They thus give a literal demonstration of how the word is made flesh. Cyrano here also argues in favor of the theory of the plurality of worlds, which envisions worlds on other planets, both in our solar system and beyond. His text thus makes for a smooth transition to my fifth chapter, where, along with Cyrano's *Lune* and *Soleil,* we will be looking at other texts that advocate this theory: the *Entretiens sur la pluralité des mondes* by Bernard le Bovier de Fontenelle, *The Discovery of a World in the Moone* by John Wilkins and Christiaan Huygens' *Cosmotheoros.*

More than on specific attempts to imagine other worlds, I will here be focusing on how metaphorical images serve as structures for thought, particularly, how cosmological systems are represented as meaningful structures by means of certain recurrent metaphors. They revolve around two basic systems. The first system includes political metaphors referring to the universe or parts of it as a conglomeration of neighboring political states or a monarchic state and its hierarchically organized society. The second one includes the metaphorical representation of the human body as a miniature cosmos or, presented from the opposite perspective, the representation of the planets and stars as human bodies. Here metaphor exceeds the purpose of mere illustration. By dictating structure and furnishing causal connections, it offers narrative coherence to

---

[42] Claudia Brodsky Lacour offers a fascinating discussion of how Descartes's 'method' in the *Discours de la méthode* is made to appear self-evident through the contextualizing autobiographical discourse of a "fable" narrated "as in a *tableau*". Focusing on the persistent use Descartes makes of architectural metaphors in the *Discours,* she shows how the self-manifestation of his method is placed in analogy to the "autonomous, self-circumscribing architectonic design", which comes to stand as a "counterdesign" to discourse: "Not a 'tableau,' 'story,' 'fable,' or any other form of representation, it is a structure in which conception is instead identical with manifestation." Cf. *Lines of Thought: Discourse, Architectonics, and the Origin of Modern Philosophy* (Durham: Duke University Press, 1996), 30 and 37. Adaptation mine. Although Descartes' rhetoric has nothing to do with visualization (indeed, its primary purpose is to avoid visual imagery by all means), what is interesting about Brodsky's argument in our context is the way Descartes' method acquires the status of a potential in the text through the discourse that contextualizes it and to which it is opposed. But see also Pierre Alain Calmé, who has pointed out that many of Descartes' images, long thought to be similes providing illustration, are actually heuristic tools that give birth to concepts later presented in mathematical form. According to Calmé, Descartes disguises the fact that the metaphor is above all a method of investigation, not just a proof of exposition. Cf. *Un autre Descartes: le philosophe et son langage* (Paris: Vrin, 1980). Discussed by Carr, *Resilience*, 2. See also Timothy Reiss, "Espaces," for Galileo's practice of using raw experience as metaphoric *vehicle* for his scientific theories; experience thus becomes the paraphrase of theory rather than being its touchstone.

explain and at times even warrant the theories proposed. While such parasitic uses of metaphors fabricate a peculiarly solipsistic picture of the universe and remain caught in topical meaning, they nevertheless offer helpful frameworks for the imagination.

Moreover, there are moments of break-down where the metaphoric structure becomes confusing and images start to blur. Such moments reveal a potential in metaphor that is truly creative and goes beyond mere bricolage. Metaphor transcends the scope of its constituting images and points to a potential beyond them. It is here that the most radically creative power of the imagination lies. In order to trace it, we need to rid the transferred metaphorical images of their pictorial quality and consider them only as frameworks of relations, structures for thought. We need to get down to the grammar of imagination, so to speak. As the very word 'imagination' implies, this is as impracticable as it is unimaginable, so that this creative potential of the imagination necessarily remains virtual. It can only exist in theory, or may whimsically manifest itself where descriptive metaphors are at war with each other.

The virtual potential of representation is also the central issue of the final chapter, dedicated to two works by Italo Calvino: *Il castello dei destini incrociati* and *Le città invisibili.* In this last chapter we are going to leave both the seventeenth century and cosmology behind to turn to ways of dealing with the limits of imaginability in contemporary literary texts. Accordingly, although we will still be focusing on ways of representing what is beyond imagination, the notion of unimaginability is here defined differently. In the preceding chapters it resulted from the dependence of the imagination on sense perception and the consequent impossibility of imagining genuine alternatives to the terrestrial world rather than mere permutations of common experiences. Unimaginability was, so to speak, a metaphysical problem[43] and, therefore, a problem of the collective rather than the individual mind. By way of contrast, in our discussion of Calvino the notion of the 'unimaginable' implies the boundless multiplicity of collective world experience and collective imaginability, and their unimaginability to the imagination of an individual.

In addition to the problem of unimaginability there is the even greater problem of the *representation* of experience and the imaginable (both individual and collective) – not just its representation in the mind, but its literary representation and narrative communicability. As we will see, Calvino's discussion of the idea of an 'encyclopedic novel' and his own attempts to create texts that appear encyclopedic, envision the idea of doing justice to the infinite universe of potentiality – a universe that either lies beyond the scope of actual representation of a text or simply beyond the scope of an author's imagination (or that of any individual). In both of the Calvino texts discussed here, the narrative constantly focuses on its omissions, thus marking a potential space for alternatives to what is said. What the text states works as a stand-in for all that it omits, so that we could call Calvino's texts potentially encyclopedic.

The change of focus from science to literature and from the narratives of the past to contemporary ones serves to emphasize that the issues relating to the imagination, its limitations and creative means, are basically the same both in science and literature and that they are still relevant today. They are certainly not an exclusive problem of seventeenth-century astronomy and cosmology. Yet as I explained at the beginning of

---

[43] Cf. p. 11, *supra*.

this introduction, the particular situation of cosmology and astronomy in the seventeenth century and its impact on the imagination makes this period in the history of cosmology and the new theories formulated in this time especially suitable for a study of imaginative representations and strategies adopted to represent what is beyond imagination. The scientific texts of this period are also particularly interesting for such a study because they are not written in a scientific language reserved for specialists. Not only did such a language not yet exist at the time, but by writing in a common language rather than Latin, many of the advocates of the 'new science' clearly, and often explicitly, targeted a wider audience (here Kepler's *Somnium* is an exception).[44]

Rather than writing for a closed group of initiates, as the Scholastics had done, these authors tried to establish communication with outsiders. In their attempt to win supporters, they were inclusive rather than exclusive.[45] This explains the great importance attached to rhetoric. Accordingly, several of these texts are written in an ostensibly literary style and in forms that imitate literary genres both classical and modern: good examples are Kepler's use of the dream vision, the use of the dialogue form in Galileo's *Dialogo*, which clearly imitates the elegant conversational style of Boccaccio's *brigata leggiadra*, and Fontenelle's adoption of the fashionable style of contemporary romances. Finally, along with their literary style, the main texts chosen for this study manifest a self-consciousness regarding their rhetoric that makes them particularly interesting for my purposes.

One could counter that the literary style and the self-critical stance of the early modern astronomical texts examined here has little in common with the specialist discourse practiced among scientists today. Yet there is a central factor that brings the discourses of poetry and contemporary science more closely together and invites their comparison. It is their dependence on a common language and their use of metaphors drawn from a shared cognitive environment as a medium – or should we rather say *corpus* – of communication. Indeed, this differentiation between language, especially metaphorical language, as transparent medium and language as *incorporated* representation is characteristic of the different positions regarding the role of metaphorical language in scientific and poetic discourse.

---

[44] For an account of the experience of a linguistic crisis among natural philosophers in the seventeenth century, see Barbara Shapiro's study *Probability and Certainty in Seventeenth-Century England* (Princeton: Princeton University Press, 1983). Shapiro offers an interesting discussion of the debates around the need to establish a neutral scientific language, a language cleansed of metaphor, ambiguity and poetic ornament in which bare facts would be allowed to speak for themselves. One of the most fervent promoters of this project was John Wilkins, a popularizer of Copernican cosmology and the author of an *Essay Toward a Real Character and a Philosophical Language* (London, 1668). However, my analysis of the central role of rhetoric in his writings will give evidence of the wide gulf between theory and practice.

[45] See, for instance, Steven Shapin's discussion of Robert Boyle's preference of a "prolix" style over a specialist mathematical language (the only specialist language available at the time) in order to target a wider audience: "It was, Boyle said, 'out of choice that I declin'd that close and concise way of writing.' He was writing not 'to credit myself, but to instruct others'." Cf. "Robert Boyle and Mathematics: Reality, Representation, and Experimental Practice," in *Science in Context* 2, no. 1 (1988), 42. See also *idem*, "Pump and Circumstance: Robert Boyle's Literary Technology," in *Social Studies of Science* 14 (1984): 481-520.

The debate around the legitimacy and specific function of metaphor in scientific discourse has a long and complex history, which I will only enter into here to the extent concerning our context. From a positivist-deductivist position, "[s]cientific language is always supposed to be literal, logical, precise, and directly verifiable."[46] Its immediate reference to reality ascertains its truth-value. According to this view, metaphors have no truth-value and therefore no cognitive value.[47] Science must either "be free of figurative language or it must control figurative language by limiting its influence so as to keep it outside of epistemology and ontology, outside of actual theories."[48] Metaphors in scientific discourse may have an exegetical or pedagogical function, but their meaning can always be paraphrased in a literal statement. The result of such paraphrasing may be extremely long-winded; to avoid this, metaphor offers a short-cut.

In contrast, revisionist views of metaphor and scientific discourse have emphasized the cognitive value of metaphor and its relation to truth, arguing that what metaphor proposes is both new and referential. This position has departed largely from the theory of metaphor introduced by I. A. Richards in the 1930s and Max Black's interaction theory of metaphor, which built on Richards's argument.[49] The basic tenet of the interaction view is that the secondary subject in a metaphor (what Richards called *vehicle*) organizes one's thought about the primary subject (Richards's *tenor*) in a new way and that this operation makes metaphor irreducible to any literal paraphrase. What happens in metaphor is an "intercourse of thoughts, as opposed to a mere shifting of words or crude substitution."[50] In adopting the interaction view of metaphor within the context

---

46 Robert Hoffman's summary of this position in "Some Implications of Metaphor for Philosophy and Psychology of Science," in Wolf Paprotté and René Dirven, eds., *The Ubiquity of Metaphor: Metaphor in Language and Thought* (Amsterdam: John Benjamins, 1985), 337.

47 For the connection between scientific realism, the positivist theory of knowledge, and the literalist view of language, see Michael Arbib and Mary Hesse, *The Construction of Reality* (Cambridge: Cambridge University Press, 1986), 147-50.

48 Hoffman, "Implications," 337. Hoffman refers to the work of Paul de Man and Mark Johnson (see bibliography).

49 I. A. Richards, *The Philosophy of Rhetoric* (Oxford: Oxford University Press, 1936); Max Black, "Metaphor," in *Models and Metaphors* (Ithaca: Cornell University Press, 1962). Some seminal works on the function of metaphor in science are: Mary Hesse, *Models and Analogies in Science* (Notre Dame: Notre Dame University Press, 1966); Arbib and Hesse, *Construction*; W. M. Leatherdale, *The Role of Analogy, Model and Metaphor in Science* (Amsterdam: North-Holland, 1974); Earl MacCormac, *Metaphor and Myth in Science and Religion* (Durham: Duke University Press, 1976); Richard Boyd, "Metaphor and Theory Change," in Andrew Ortony, ed., *Metaphor and Thought* (Cambridge: Cambridge University Press, 1979), 356-408; and, in answer to Boyd, Thomas Kuhn, "Metaphor in Science," *ibid.*, 409-19.

50 Janet Martin Soskice and Rom Harré, "Metaphor in Science," in Zdravko Radman, ed., *From a Metaphorical Point of View: A Multidisciplinary Approach to the Cognitive Content of Metaphor* (Berlin: de Gruyter, 1995), 294. I am here only interested in the basic concept of Richards's and Black's reevaluation of metaphor in the abbreviated form in which it proved influential in revisionist work on scientific metaphor. Therefore, I will neither go into the differences between their theories, nor into Black's reformulation of his own theory in "More about Metaphor," (in Ortony, *Metaphor*, 19-43), which merely weakened his original approach without solving its problems.

of science, theorists have emphasized the heuristic function of scientific metaphor and its ability to make new meaning intelligible:

> Only by some such device as metaphor is it possible to develop new meanings in new theories that are intelligible. Metaphors have a hypothetical nature; they suggest possible new meanings to us.[51]

Yet even among the still relatively small number of theorists who have argued for the constitutive nature of language and metaphor there is often disagreement "as to what it *means* to say that language, or metaphor, is constitutive of science." In acknowledging the cognitive value of metaphoric meaning, many revisionists are concerned that metaphors might get out of hand. As James Bono points out, some among them therefore propose a "domesticated" role of metaphor in science by claiming that the privileged insight of scientists controls the unruliness of metaphoric meaning.[52] This attempt to delimit proliferation of meaning in metaphor is particularly apparent where theorists contrast their definitions of the nature and function of scientific metaphors with those of literary metaphors.

Literary metaphors are said to differ from scientific metaphors in their "striking", "shocking" and "often formally contradictory" nature,[53] in their "greater diffuseness", their "emotive and affective tone";[54] they typically lack concern with objective reality[55] and they need to be novel in order to be effective.[56] In contrast, scientific metaphors are working tools: they will be shared and developed by a large community of scientists if they prove fruitful. According to Richard Boyd, both literary and scientific metaphor are open-ended, but in different ways: Literary metaphor is 'conceptually open-ended', which is to say that "the function of literary metaphor is not typically to send the informed reader on a research project;" rather, such a reader is expected to be already sufficiently informed to understand the metaphor. Scientific metaphors, on the other hand, are research-oriented and therefore open for inductive development. Ultimately, however, explication is the aim.[57] Even much more radical theorists like Michael Arbib and Mary Hesse, who refer to scientific metaphors as "freely imagined narratives about the natural world" and who stress the ideological component in such narratives, emphasize that metaphors in science are "intended to satisfy ... the

---

[51] MacCormac, *Metaphor and Myth*, 36.

[52] "Science, Discourse, and Literature: The Role/Rule of Metaphor in Science," in Stuart Peterfreund, ed., *Literature and Science: Theory and Practice* (Boston: Northwestern University Press, 1990), 60-61. Similarly, Gillian Beer discusses the tendency of professional groups "to exclude, or suppress, feasible meanings". She argues that such meanings remain potential. "They can be brought to the surface and put to use by those outside the accord or professional 'contract', as well as by those future readers for whom new historical sequences have intervened." Cf. "Problems of Description in the Language of Discovery," in George Levine, ed., *One Culture: Essays in Science and Literature* (Madison: University of Wisconsin Press, 1987), 42.

[53] Arbib and Hesse, *Construction*, 157.

[54] W. H. Leatherdale, *Role of Analogy*, quoted in Bono, "Role/Rule," 67. I am much indebted here to Bono's critical discussion of approaches.

[55] *Ibid.*

[56] Boyd, "Metaphor," 362-64.

[57] *Ibid.*

pragmatic criterion ... by being constrained by feedback loops involving experimentation in the natural world."[58]

It is not my intention to deny that there are functional differences between scientific and literary metaphors, or that science must refer more immediately to the real world than literature. Yet the focus in the arguments referred to above is too much on the pragmatic function of metaphor in as much as they represent metaphor as a mere tool to be discarded as soon as the metaphorical meaning interferes with any empirical facts of science. These theorists allow for metaphoric proliferation of meaning only as long as the metaphor is "fruitful". They thus present metaphor as a *temporary process*, to be superseded by explication or dismissal. This is done by means of an all too crude distinction between literary and scientific discourse that relies on a "caricature of literature and literary creativity".[59]

Bono rightly criticizes Boyd's comparative account of literary and scientific metaphor for attributing the unruliness of literary metaphor to the subjective, personal choice of an author, rather than to language itself.[60] This concept of literary language as a highly personal, idiosyncratic language is also at work in the attempt to establish a privileged access of science to the real world by distinguishing science from the 'lack of concern with objective reality' attributed to literary discourse. While I agree that literature does not claim to represent 'objective reality' (such a claim would appear highly dubious also with regard to science), I think it is important to emphasize that literary language and, more specifically, literary metaphor do refer to a shared world of experience – indeed, that they must, if they are to be meaningful at all. Although metaphorical meaning may be open-ended, the informed reader (as Boyd himself suggests) can be expected to grasp its basic drift without having to embark on a research project. Despite radical attempts by post-structuralist and deconstructivist critics to do away with the concept of authorship altogether, understanding metaphor – like irony or parody – does rely on some notion of authorial intention and its reconstruction on the part of the reader or audience, even if such a reconstruction will always vary in accordance with the reader's cultural and historical background. Metaphoric meaning, like all meaning, relies on intersubjective understanding, on a shared cultural code and, therefore, on some kind of guidance as to how a metaphor is to be understood. This is true for current metaphors as well as for new, 'striking' and 'shocking' ones.

In science, where one of the central functions of metaphor is that of "represent[ing] the unknown in terms of the known",[61] the tenor of the metaphor (or what Mary Hesse, with specific reference to a scientific context, terms the "domain of the explanandum")[62] is often inaccessible to experience. One could therefore argue that the structural framework suggested by the metaphor has a more unidirectional and manipulative influence on the scientist's perception of the explanandum than in the case of literary metaphors, where the reader's notions of tenor and vehicle interact with each other. In

58 *Construction*, 157-59.

59 Bono, "Role/Rule," 65. Bono is referring to Boyd's distinction alone, but I think this also applies to Leatherdale and even, to some extent, to the argument presented by Arbib and Hesse.

60 *Ibid.*, 65-66.

61 MacCormac, *Metaphor and Myth*, 140.

62 Hesse, *Models*, 159. The term Hesse assigns to the metaphoric vehicle is "explanans".

scientific metaphor, interaction between the domain of the explanans and the domain of the explanandum can only take place as a result of experimentation and the feedback loops it furnishes. It is therefore always to some extent delayed. I am saying 'to some extent', because I agree with Hesse that some antecedent similarity or analogy must be discerned between the metaphor and the explanandum if the metaphor is not to be wholly arbitrary.[63] However, I do not agree with her categorical rejection in the context of science of Black's suggestion that metaphor "creates the similarity [between primary and secondary subject]" rather than "formulates some similarity antecedently existing".[64] In the process of the development of the metaphor, the creation of fictional similarities plays an important role in hypothesis building.

To sum up: It is important to acknowledge the fact that scientific metaphors are not neutral frameworks of scientific investigation, that they remain grounded in the sociocultural and political context from which they are taken and that the kind of proliferation and dissemination of meaning more readily associated with literary metaphors is also at work in scientific metaphor. As Gillian Beer has pointed out, the "free and multiple movement across levels" typical of literary language "is its characteristic resource of discovery."[65] It is this resource that is exploited in scientific metaphor. Moreover, old metaphors that have been overruled by new findings (the feedback loops of empirical investigation) often reappear in arguments with which they are quite incompatible. The reality posited by science is both shaped and understood through metaphor and the contexts it invokes. This is certainly more easily noticeable in early modern scientific texts than in contemporary ones. The reason for this is not just that the former are written in a much more obviously literary style, but also that it is easier to assume a position of critical distance in relation to cultural paradigms that no longer coincide with our own.[66]

Science strives to increase knowledge and is therefore wary of acknowledging its own limits or any absolute limits of human knowledge (such as the eternal questions regarding the origins of the cosmos or the existence of God). It does not like to lay bare the fictions involved in its mechanisms, but pretends "to eliminate from language all uncontrolled elements and to turn language into a tool which lacks any will of its own; a tool that subjects itself freely and entirely to thought."[67] As a form of writing that is not concerned with progress, literature – as opposed to science – is at much

---

63 *Ibid.*, 161-62.

64 *Ibid.*, 37. Addition mine.

65 "Problems," 49.

66 Interesting work on the powerful influence of cultural stereotypes on scientific investigation has been presented in Ludmilla Jordanova's *Sexual Visions: Images of Gender in Science and Medicine between the Eighteenth and Twentieth Century* (New York: Harvester Wheatsheaf, 1989) as well as in Nancy Leys Stepan's study of the role of race and gender in the eugenics movement in Latin America, the United States and Europe: *"The Hour of Eugenics:" Race, Gender, and Nation in Latin America* (Ithaca: Cornell University Press, 1991). See also Ludmilla Jordanova, ed., *Languages of Nature: Critical Essays on Science and Literature* (London: Free Association Books, 1986), which focuses on the emphasis on stereotypical notions of sexuality, gender and kinship in the perception of nature, especially human nature.

67 Hans-Jost Frey, *Die Autorität der Sprache* (Lana: Edition Howeg, 1999), 18. My translation. Frey speaks of an "attempt" on the part of science; I refer to it as a 'pretense' because scientific discourse often works by more oblique means than it acknowledges.

greater liberty to acknowledge its own limitations. In fact, more than just acknowledging limitations, it plays with them and, in this way, draws attention to them. Literary texts surrender themselves to the authority of language. By approaching scientific texts with methods that owe more to literary criticism than to the history of science as traditionally conceived, I hope to offer a more critical assessment of techniques of representation and the use of rhetoric in science. Such an approach may sharpen our awareness of the fact that scientific argumentation often depends on hypothetical frameworks and, at times, on fictions with no empirical foundation whatsoever. At the same time, it is in the scientific race for progress that metaphors serve as a tool for investigation, which, however illegitimate, carries a heuristic potential by shaping the paradigms in which we think. There is, of course, always the very real danger that metaphors deceive, and it is not restricted to science and literature. Nevertheless, the use of metaphors as cognitive frameworks most clearly reveals the innovative and creative power of language and the imagination.

As a kind of metanarrative and poetics for this study of the creative potential of the imagination, I am going to refer to the episode on the Moon in Ariosto's *Orlando furioso*. I suggest we read this fantastic lunar voyage as an allegory on the human need to appropriate the strange, on the limitations of the imagination, and, finally, on poetic creativity. It thus combines the key issues of my analysis of astronomical texts with the broader issues regarding the creativity of the imagination that I attempt to raise here.

Ariosto's lunar fantasy moreover fits in as a prelude to our discussion of Galileo and Kepler, for his sources included Plutarch's *De facie in orbe lunae*, which became an influential source for both astronomers. I have mentioned earlier that the Moon is represented in this dialogue as a kind of counter-Earth, which is why Plutarch later became an important authority for defenders of Copernicanism. That Ariosto should have alluded to such an unorthodox text even before the introduction of Copernican theory may appear striking and may make him look like a cosmological revolutionary. Yet this is hardly the case. In Ariosto, the Moon is still an *outopos*, a 'no-place', which reflects life on Earth with an ironic distance. In doing so, his text plays with the notion of a change of perspective that is fundamental both to Copernican theory and to imagining a strange universe. Finally, starting a discussion of the role and significance of the imagination in astronomical writings with Ariosto's more poetic concerns will facilitate the transition to Calvino and to considerations regarding the imagination in more overtly literary texts.

1

How Metaphors Matter: Astolfo's Lunar Journey in the *Orlando furioso*

The scene is set on the Moon, the story is one of lunacy. Orlando, the greatest warrior in Charlemagne's army, has been punished by God with madness for deserting the war against the Saracens in amorous pursuit of the heathen princess Angelica. Now the fixed period God ordained for his ravings is past and Orlando's wits must be recaptured if the Christians are to win the holy war. This task God has assigned to the English knight Astolfo: he is to reconquer Orlando's wit on a journey to the Moon, where all the things lost on Earth end up. Before taking off, Astolfo flies with his winged steed to the terrestrial paradise, which lies on top of a high mountain, close to the source of the Nile. He is welcomed there by John the Evangelist, who presents himself as a travel guide to Astolfo, and the next morning the two travelers depart on their journey.

On their arrival in the world of the Moon, Astolfo is struck by the vastness of the Moon and the insignificant size of the Earth.[1] Once he starts looking around, he sees

> Altri fiumi, altri laghi, altre campagne
> sono là su, che non son qui tra noi;
> altri piani, altre valli, altre montagne,
> c'han le cittadi, hanno i castelli suoi,
> con case de le quai mai le più magne
> non vide il paladin prima né poi:
> e vi sono ample e solitarie selve,
> ove le ninfe ognor cacciano belve.[2]

> Other rivers, other lakes, other pastures
> are up there that are not found on Earth;
> other plains, other valleys, other mountains,
> they have their own cities and their own castles,
> have houses of such a big size as the
> paladin had never seen before, and would never see again:
> and there are ample and solitary woods
> where the nymphs are engaged with catching wild beasts.[3]

---

1 The observation regarding the relativity of size between Earth and Moon acquired great significance in the context of Copernicanism, in which the Earth was turned into a planet, and, especially, the theory of the plurality of worlds, which developed out of it. It is noticeable that such an observation should be made in a text predating Copernicus' *De Revolutionibus Orbium Coelestium* (1543). The same observation can, however, already be found in Lucian's *Ikaromenippos*, which served Ariosto as a source.

2 Ludovico Ariosto, *Orlando furioso*, ed. Cesare Segre (Milan: Mondadori, 1976), xxxiv, 72. All quotations in Italian refer to this edition.

3 I have made my own verbatim translation of this passage since the English translation by John Harington that I use elsewhere departs here in essential ways from the original and leaves out half of the stanza.

Strangely enough, Astolfo does not in the least wonder at the incredible similarity this place bears to the world he comes from. This is ironic, for the rivers, lakes, cities and castles – though apparently 'other', different from the ones he knows from the Earth – are still recognizable as rivers, lakes, cities and castles. The narrator makes this clear by insisting on the word 'altri'/other, so as to emphasize the essential *analogy* between the two worlds. As Georges Güntert has pointed out in a lecture on the *Orlando furioso*, what is "typical of the entire presentation of Ariosto's lunar world is that the concepts of diversity and analogy (non-diversity) coexist without standing in a relation of opposition to each other."[4]

Continuing their journey to a narrow valley, Astolfo and John find the place cluttered with matter, for all that was lost on Earth – be it through human shortcomings, time or fate – has resurfaced on the Moon. This may be an ironic comment on the traditional Aristotelian tenet of the purity and smoothness of the lunar globe, a tenet that Galileo's telescopic discoveries were to disprove. In fact, immediately upon his arrival, Astolfo succumbs to the illusion of seeing the Moon in its traditional guise, "come un acciar che non ha macchia alcuna" (xxxiv, 70,4), which John Harington translates with some addition: "The Moone was like a glasse all voyde of spot/Or like a peece of purelie burnisht steele."[5] Astolfo, it seems, is a brainwashed victim of Aristotelian dogma, since he is able to recognize only at second sight that the Moon's surface does in fact look quite different from what he expected it to be: it is the Earth's dump.[6]

Without the help of John the Evangelist, who serves as an interpreter, Astolfo would be quite at a loss as to the significance of the great heaps of matter on the Moon. Among other things, John explains to him that the heaps of burst crickets are the verses rehearsed in praise of great lords, that the large and hard-blowing bellows are the favors which princes grant to their Ganimedes and which rapidly dwindle away as the young man's beauty wanes, that spilt soup stands for deeds of charity that came too late, while a great number of snares covered with bird-lime represent the ladies' lost beauty.

Contrary to the apparent *analogy* between the two worlds sustained before, this passage at first suggests a relation of *complementarity* between them: what disappeared from the Earth, has resurfaced on the Moon. And yet on closer inspection, this relation too turns out to be incorrect since what disappeared from the Earth does not reappear on the Moon in a form identical to its earthly one; instead, the heaps of matter on the Moon stand in a *metaphorical* relation to terrestrial losses. As the Evangelist explains, this also applies to the larger system of correspondences between Earth and heaven

---

4 Cf. *L'Orlando furioso*. Script of lecture (University of Zürich, summer 1993). Translation from the Italian mine.

5 Ariosto, *Orlando furioso: Translated into English Heroical Verse by Sir John Harington* (1591), ed. Robert McNulty (Oxford: Clarendon Press, 1972), xxxiv, 70, 5-6.

6 Ariosto thus brings into play two diametrically opposed, though equally current, symbolic readings of the Moon: the Moon as a symbol of purity (which was often used as an icon for the Virgin, especially in her guise as Immacolata) and the Apocalyptic Moon, a symbol of corruption and inconstancy. Cf. Steven Ostrow. "Cigoli's Immacolata and Galileo's Moon: Astronomy and the Virgin in Early Seicento Rome," *Art Bulletin* 77, no. 2 (1996): 218-35; see also idem, *Art and Spirituality in Counter-Reformation Rome* (Cambridge: Cambridge University Press, 1996), 243.

which rules that every single incident on Earth find its correspondence in heaven in a different guise:

– Tu déi saper che non si muove fronda là giù, che segno qui non se ne faccia. Ogni effetto convien che corrisponda in terra e in ciel, ma con diversa faccia. (xxxv, 18, 1-4)	Know first (said he) there cannot wag a straw Below on Earth but that the signe is here, And each small act doth correspondence draw Although in other shew it doth appeare.[7]

As it turns out, Astolfo is incapable of understanding the significance of these metaphors without the help of his interpreter. This is ironic, for metaphors such as snares covered with bird-lime standing for female beauty or blowing bellows representing courtly hypocrisy are highly conventional. Astolfo's incomprehension allows for two different interpretations: It may, first of all, be taken as a comment on the inability characteristic of human beings to see the world around them in its true colors. According to Enrico Musacchio, the difference between the lunar and the terrestrial world is that on the Moon everything is relentlessly shown as what it is in its crudest reality.[8]

Secondly, Astolfo's helplessness in understanding the lunar metaphors questions the legitimacy of these metaphors and of poetic conventions more generally: after all, what else but a poetic convention determines that snares covered with bird-lime signify a lady's beauty? The fact that Astolfo is unable to understand the meaning of even the most conventional metaphors shows how arbitrary they become once they are ripped out of their familiar context.

The arbitrariness of metaphoric representation is quite shockingly exposed where one would least expect it: in the allegory on fame, which follows Astolfo's visit to the valley of lost things. When Father Time throws the name plates of the deceased into the river of forgetfulness, a swarm of vultures and crows try in vain to preserve some of the most exquisitely adorned plates from eternal oblivion. Overwhelmed with the weight of their burden, the birds return it to the rushing stream. Only two white swans – suspiciously as white as the eagle in Ippolito d'Este's (Ariosto's employer's) coat of arms – succeed in rescuing the choicest of names. These are preserved for eternity in the Temple of Immortality. According to St John, the vultures and crows are the bootlickers and flatterers that frequent the courts of the great in search of a comfortable life, but who will spit out their benefactor's name when he or she is no longer useful to them. The beautiful swans, on the other hand, are the poets in whose works their patrons win eternal fame. The aptness of this representation is, however, soon qualified when we find out that the swans pick from the stream only the names of

7 Harrington's translation is vague here: in the Italian version this rule does not refer to the Moon alone, but to the heavens at large.

8 *Amore, ragione e follia: Una rilettura dell'Orlando Furioso* (Rome: Bulzoni, 1983), 216. This interpretation also corresponds to a point Peter DeSa Wiggins makes regarding Orlando's character: Wiggins thinks that even before Orlando loses his wits, an inability to recognize characteristic traits in others as well as an inability to assess his own situation is latent in his character. In the context of Ariosto's lunar allegory, this trait in Orlando may well reflect a general human characteristic. Cf. *Figures in Ariosto's Tapestry: Character and Design in the Orlando furioso* (Baltimore: Johns Hopkins University Press, 1986), 109-39.

the potentates who were generous patrons of poets. This is, of course, an implicit attack on Ippolito d'Este, under whose neglect Ariosto suffered during a great part of his life; but it also makes a statement about poetic misrepresentation of reality:

Omero Agamennón vittorïoso,
e fe' i Troian parer vili et inerti;
e che Penelopea fida al suo sposo
dai Prochi mille oltraggi avea sofferti.
E se tu vuoi che 'l ver non ti sia ascoso,
tutta al contrario l'istoria converti:
che i Greci rotti, e che Troia vittrice,
e che Penelopea fu meretrice.
(OF, xxxv, 27)

Homer made Agamemnon victorious
and made the Trojans appear vile and sluggish,
said that Penelope remained faithful to her husband
and suffered a thousand wrongs from her suitors.
Yet if you want to know the truth,
you have to turn the whole story upside down:
that the Greeks were beaten, that the Trojans won,
and that Penelope was a common whore.[9]

Similarly, the Roman emperor Augustus was not as great a saint as Virgil made him appear while, on the other hand, Nero would enjoy a better reputation, had he been better at making friends with writers. The poets, so augustly portrayed in the allegory, thus fall into line with the courtly sycophants; the swans join the vultures and crows. The indictment against poets and historians for being venal liars implicitly even falls back on John the Evangelist himself, the quintessential writer, as it were: He explains his sympathy for writers by drawing attention to the fact that he himself used to be one them, and then praises Christ, his patron, for the generous reward he offered him for his services. Yet despite all the lying and fawning, these great names in Western culture are here to stay; their fame has been eternalized in the topography of the Moon.

Granted that what the Evangelist says is true, this has an important bearing on the status of the Moon. It implies that, rather than a skewed mirror which presents life on Earth 'relentlessly in its crudest reality', as Musacchio argues, the Moon is a faithful mirror which faithfully renders all the terrestrial lies.

Since, as he only too frequently implies, the author of the *Orlando furioso* is equally guilty of the kind of misrepresentations customary in his trade, we must be wary of his tale. We cannot take it at face value. At the same time, we should be careful about endorsing the interpretation the narrator offers us. Let me propose instead that we read this allegorical scene on the Moon as yet another allegory, one that concerns itself with the conventions of allegory and hence with the legitimacy of allegorical and metaphorical representation as such. In doing so, it probes the very foundations of imaginative fictions, raising questions regarding the creative potential of the imagination and its limitations.

The spotless Moon, which Astolfo perceives at the very outset of the journey, is the blank page – it is less; as a place so remote and utterly strange, it is the emblem of complete objectivity, of a true beginning. At the same time, it is already a projection: The Moon, seen with the naked eye, is not spotless. And the projections continue: The world the narrator evokes from the blank page is said to record terrestrial occurrences in minute detail, though 'with a different face',[10] so as to manifest the divine correspondence between heaven and Earth. The 'different' face of terrestrial reality is, one would expect, its *true* face, the transcendental image of history that corresponds to

[9] The translation is mine.

[10] "ma con diversa faccia" as quoted on p. 25, *supra*.

the divine plan. Instead, every aspect of terrestrial life has ironically turned into matter and appears in the most conventional of allegoric imagery: flocks of linen, wool, cotton or silk signify the lives of human beings; some ugly, some beautiful in color, depending on the person whose life they represent. The Three Fates spin these flocks to threads and weave them into ornamental tapestries;[11] to each individual thread corresponds a plate, made either of iron, silver or gold, which carries the name of each person. The plates are then thrown into the river of forgetfulness by Time, a nimble old bearded man committed to destruction, and only a small number of them are rescued from the flood by two beautiful swans and taken to the temple of Immortality, where they are hung by a beautiful nymph, Fame.

One might object that the conventionality of the imagery is a necessary characteristic of the genre of allegory. And this is, of course, true. Yet it is also just the point. Without the element of convention, the allegory would remain incomprehensible. Ariosto's allegorical lunar world therefore in many ways resembles the kind of allegorical representation typical of the epic poetry of the period to the point of being interchangeable with it. There is, however, a moment of distancing that marks the passage from solemn allegory to ironic mimicry. It occurs with the self-indictment of the poets: The Evangelist's affirmation that human fame entirely depends on the fictions fabricated by poets turns the re-enactment of these fictions in the lunar reflection of earthly history into a poetic fabrication. The executors of divine Providence in the allegory have, as it were, gotten involved in a scheme of corruption concocted by the poets.

Here the allegory turns on itself and draws attention to its mode of representation. Allegory turns into anti-allegory so as to imply that the imagery *should* really be incomprehensible because it has nothing to do with truth. As a consequence, the allegorical image loses its power to signify by proxy and is reduced to its intrinsic material value, the stuff it is made of: old spinning women, an old man, a river, some birds, a beautiful nymph, a temple with ludicrous name-plates. In Ariosto's allegory or anti-allegory of allegorical representation it is *only* convention which allows these items to become the image of transcendental truth or of any reality beyond their own.

That such reflections should themselves figure in the image of allegory may seem ironic, yet it lends the argument more poignancy by offering a complementary illustration of its content. The ironical distance with which this allegory presents all allegorical representation including itself reflects in a wider sense on imaginary creations of virtual (or non-perceptual) scenes: In its creation of the virtual, the poetic imagination is essentially dependent on visual impressions from the perceptual world; any new fiction inevitably relies on the fictions of old and, ultimately, on a common experience of the material world. Hence, all representation of the virtual is essentially metaphorical in nature. It can only be realized by means of substitution with elements that belong to the familiar world of sensual perceptions. Accordingly, the imagination can only create by means of metaphor; it is metaphoric in its very nature. Viewed in this light, the

---

11 Alexandre H. Krappe notes that in a popular legend the lunar spots were said to be a female weaver exiled on the Moon for having woven on a holiday or during a period when it was forbidden. Cf. *La genèse des mythes* (Paris: Payot, 1938), 120-21. Accordingly, the poetic image of weaving, often used as a metaphor for poetic creation, is traditionally associated with the Moon.

conventionality of the allegorical imagery in the ironic representation of transcendental truth seems to be deliberately overdone so as better to convey a sense of the inevitable restriction of the poetic imagination to the material world of human experience.

This argument could not find a more literal expression than in the episode recounting the visit to the valley of lost things, where the metaphorical letter has quite literally turned into matter. Stuffed with heaps of matter, the valley evokes a waste-yard of useless things. Yet the substantialness of these metaphorical signifiers stands in marked contrast to their meaning: Much of what has disappeared from the Earth and has resurfaced on the Moon has to do with human vanity, with people's vain hopes and desires, or is simply insubstantial. It bears a relation to vacuity, emptiness, nothingness and delusion, to 'things' that either have no matter or that do not matter. On the Moon, paradoxically, all this vacuity has materialized and turned into objects intended to express its true nature.

By far the greatest heap of matter that Astolfo finds there consists of bottles filled with an exquisite liquor representing human wit. Indeed, this is no longer a heap, it has grown to a veritable mountain. The various bottles are of all different sizes, each labeled with the name of the former carrier of its contents. The biggest bottle carries Orlando's name; it is the bottle Astolfo will have to take home from his journey in order to turn the raving brute back into the noble paladin Orlando. Astolfo is, moreover, allowed to take a sniff at his own bottle to the effect that for some time after he lives wisely. In this way, even a concept as completely incompatible with any substance as human reason has materialized into a liquor reminiscent of some quack's medication.

These metaphorical representations thus serve to make visible and to characterize that which has no face of its own. Through metaphor the immaterial materializes in our mind and becomes imaginable. It is only metaphorically that we can 'grasp' its significance; its image is always the image of something else. The materiality with which wit features on the Moon is thus itself a paradigmatic illustration of how the human mind imagines or wit works. Yet Astolfo's obtuseness in the face of the *nature morte* of metaphors belies their relation to what they are there to signify and, accordingly, their serviceableness in revealing the true nature of things or anything beyond their immediate appearance. When Astolfo later inhales the contents of his own bottle to regain full presence of mind, bottle and contents no longer figure as a transcendental image of wit, but have become the thing itself. Once again, metaphor loses its meaning and lasts only as matter.

From a metanarrative perspective, the use of metaphor and allegory in Ariosto's lunar episode thus offers a poetics of the imagination, of the making of images. By calling attention to the essentially metaphoric nature of imagination, it asks us to consider its creative potential as well as limits, its confinement to sensual experience in the visualization of the virtual and, more broadly, its re-employment of conventional imagery in the fabrication of new fictions. We find here an illustration of the poet's impotence in creating images except by transposition and recombination. At the same time, however, and in full compliance with this illustration, we have become witnesses of the marvelous art of imagination and poetic creativity, and this not only on this truly phenomenal lunar journey, but on our journey through the *Orlando furioso* as a whole, which is both a rehearsal of the lunar poetics *and* their enactment.

For one thing, Ariosto sets the story of the *Orlando furioso* in the remote history of Charlemagne's wars against the Saracens, thus evoking the literary traditions of the Carolingian and the Breton cycle as well as that of French medieval romances. But then this choice of setting cannot even be considered original for being unusual: It rather follows the fashion of chivalric epics highly in vogue at the time, of which Pulci and Boiardo were the most notable exponents. Indeed, Ariosto tells a story that quite obviously is not his own, at least in so far as it is a continuation of Boiardo's *Orlando innamorato*, of which he retains the main characters and their antecedents.

While the practice of adapting and elaborating old stories is constitutive of the very genre of chivalric epic, I think that Ariosto's decision to circumscribe his range of possibilities for more personal imaginative creations to such a degree and in such an obvious way is significant, especially in view of the lunar episode and the argument about literary conventions and the limitations of the imagination it implies. It allows him to demonstrate more effectively how the imagination works and in what ways it becomes creative even while depending on old and familiar material. For when Ariosto decided to write in the genre of chivalric epic and, more specifically, when he took it upon himself to continue the old tale where Boiardo had left it, he was clearly determined to outdo his predecessors by a long shot. Of this he assures us right at the beginning of his poem – notably by quoting Petrarch as well as Horace:

Dirò d'Orlando in un medesmo tratto	I will no lesse *Orlandos* acts declare
cosa non detta in prosa mai nè in rima:	(A tale in prose ne verse yet song or said)
(I, ii, 1-2, my emphasis)[12]	

Yet in fulfilling his promise Ariosto relied on the same material as his predecessors, not only in terms of setting, but of characters and theme. In fact, the promise voiced in the verses just quoted implicitly challenges Boiardo by claiming to outdo him in the principle subject matter of his poem: Orlando's state of being in love – in the *Furioso*, the still contained 'innamorato' has turned into raving love-besottedness:

che per amor venne in furore e matto,	Who fell bestraught with love, a hap most rare
d'uom che sì saggio era stimato prima;	To one that erst was counted wise and stayd.
(I, ii, 3-4)	

At the same time, the furious turn that Ariosto gives to the title of his poem is not an innovation in itself, but evokes the *Hercules furens* by Seneca. Ariosto's challenge 'to say things never yet said in prose or verse' may thus refer to the broad range of his borrowings and the daring yet masterful way in which they are combined and integrated into the work at large, rather than to more original innovations. This program finds expression in the title of the work, where Orlando, of more humble Romance origin, meets his equal in the noble Hercules.

Ariosto's literary borrowings are indeed profuse and wide-ranging: Besides the subject matter and the principal characters, which he took over from Boiardo, he

12 Cesare Segre indicates the sources in a note: "Carmina non prius Audita Musarum ... Cano", Horace, *Carmina*, III, i, 2-4; and "che nè 'n rima Poria, nè 'n prosa ornar assai", Petrarch, *Triumphus Cupidinis*, IV 70-71. (*Orlando furioso*, 1267, note 4)

adopted characters and subplots as well as stylistic elements and verbal quotations from vastly different sources, ranging from Greek and Latin authors and the great Italians (Dante, Petrarch and Boccaccio), to the Romance tradition, the *chansons de geste*, popular myths, his Italian predecessors in writings of chivalric poetry, and contemporary poets like Poliziano, who worked in quite different genres. Despite their diverse origins, all these elements are so naturally and harmoniously combined with each other as to create the impression that they could never have been used elsewhere, that they are indeed novel. They are worked into an exquisite tapestry (*la gran tela*, as Ariosto frequently refers to his poem), an intricate network of interrelated stories, which is at the same time vastly complex and rigorously controlled.

Ariosto lives up to his challenge – masterfully and wonderfully so. And a good part of the boldness of the challenge lies in the fact that the threads which make up the complex design of his tapestry have been used time and again in the composition of other stories. Yet, as the metaphor of the tapestry suggests, the art of poetic imagination lies not in *inventing* the basic material, not in creating *ex nihilo*, but in choosing from what is given, in piecing the individual elements together, in drawing the connections, in inflating and deflating, in creating distance and changing perspectives, in creating new wonderment out of old stuff. This is how the Three Fates on the Moon keep turning the past lives of women and men into art. This is how Arachne outmatched Athena. And, to return to Seneca's bee image, is not just this the 'spiritus' of the artist that turns nectar into honey?[13]

The Moon is thus not the pristine page waiting to be filled with images never seen before; instead it is the warehouse of literature full of old rubbish that can become useful again by being integrated into a new context, woven into a new story. There it can come to life again. Ariosto's poetic imagination manifests itself in the rich expression of the subtle life of relation that runs between the characters, the actions and backgrounds. As we shall see, it is not by coincidence that Italo Calvino, a great admirer of Ariosto's, defined the modern novel in almost exactly these terms.

But let us return once again to our traveler, who has been standing around nonplussed, strenuously trying to make sense of what he is seeing. With reference to Astolfo, the essential sameness of the lunar and the terrestrial world may reflect the traveler's inability to discover anything different from what he is familiar with. In an attempt to make sense of the strange new world around him, Astolfo forces it into a relation of analogy (or non-diversity) to the world he knows. The world Astolfo finds on the Moon is thus essentially similar to the terrestrial world not because it is so in reality, but because of Astolfo's limited perceptual abilities – he cannot see it differently. Accordingly, visual perception is presented as a function of the human mind that appropriates and thus domesticates everything it takes in. By analogically relating any hitherto unprecedented perception to the familiar world, the mind creates an image of the world in which the idea of difference is, at most, one of varied sameness.

The astronomical descriptions of the Moon by Galileo and Kepler, which we are going to turn to next, also translate the strange image of the Moon into familiar images so as to make it both meaningful and imaginable. The problem with such descriptions is that they do not leave room for the possibility of radical differences between the

---

[13] Cf. Introduction, 9-10, *supra*.

lunar and the terrestrial world, but represent a geomorphic view of the universe that strikes an awkward note in an argument for Copernicanism. Let us look at these descriptions in detail and see what kind of remedies they propose for such shortcomings.

2

Images Proposed in Jest: Galileo's *Sidereus nuncius* and the *Dialogue*

In the *Sidereus nuncius*, Galileo offers descriptions of the astonishing new world he discovered when for the first time he observed the heavens through his telescope. His message is the following: he begins with a description of the lunar surface, which he presents as rough and essentially Earth-like in nature. The description is based on the way in which spots on the surface of the Moon change shape over time: by drawing analogies to the Earth, Galileo identifies this phenomenon as shadows cast by mountains on the moon. Along with the discovery of a great number of stars invisible to the naked eye, Galileo offers a new explanation for the phenomenon of the Milky Way: it is an accumulation of innumerable stars. This discovery is of great importance to the defense of Copernican theory for it supports the thesis that the universe is really much vaster than previously believed and thus explains the apparent absence of stellar parallax. Galileo closes with a narrative relating his discovery of four of Jupiter's satellites. He uses this discovery as evidence for his argument that to have an orbiting moon (as the Earth does) does not inhibit planetary motion and that, therefore, both the Earth and Jupiter represent mini-models of the Copernican system within the larger one centered by the Sun.

The little book struck like a bomb: This was the first time that someone offered strong visual evidence in support of Copernican theory, which up to that point had been so little convincing to most people because of the way it contradicted sense perception and common sense. It was also the first time that someone challenged the validity of the Aristotelian description of the Moon as a perfect sphere on the ground of new visual evidence.[1] At the same time, Galileo clearly infringed on the regulations regarding the kind of claims that could legitimately be made about the celestial realm with his claims regarding the physical nature of the Moon and his way of arguing for

---

1 Galileo was not the first to propose such a lunar theory: Democritus had accounted for the spots on the lunar surface by proposing the existence of lunar hills and valleys. Terrestrial similarities in the moon had been proposed, moreover, by Anaxagoras, Pythagoras, Philolaus, and many others. Cf. Marjorie Hope Nicholson, "A World in the Moon: A Study of the Changing Attitude toward the Moon in the Seventeenth and Eighteenth Centuries," in *Smith College Studies in Modern Languages* 17, no. 2 (1936), 1. Moreover, in his *De facie in orbe lunae*, Plutarch had presented various theories on the nature of the Moon circulating at the time; yet all of these had remained unfounded or had only been based on the spots that are visible with the naked eye. Cf. "Concerning the Face which Appears in the Orb of the Moon," in *Moralia*, trans. Harold Cherniss (Cambridge, MA: Harvard University Press, 1957), vol. 12.

them by analogically relating visual observations to terrestrial phenomena. His claims, moreover, constituted a challenge to the status traditionally assigned to astronomy in the hierarchy of disciplines.[2]

Ezio Raimondi points out that Galileo conferred onto vision an essential classifying function. The act of seeing rose to the position of an intermediary operation between the process of sense perception and that of the intellectual elaboration and reconstruction in the mind so that visual experience for the first time coincided with cognitive experience.[3] However, in order to allow for conclusions regarding the qualitative nature of the Moon (or the universe more generally) to be drawn on the basis of visual evidence, one first needs to establish the premise that there exists a substantial relation between the Earth and other celestial bodies. Yet this premise contradicts the fundamental principle of Aristotelian cosmology, which distinguishes between a terrestrial and a celestial realm and holds that these realms are made up of radically different substances. According to Aristotelian doctrine, which was the position firmly held by the philosophers at the universities and supported by the Church[4], all that could be said about celestial bodies was that they were eternal, unchanging and perfect in nature and in the kind of movement that pertained to their nature. In this they radically differed from anything observed on Earth. It was therefore absolutely inadmissible to derive qualitative assumptions about the celestial bodies from any laws or observations that concerned the Earth.

If any qualitative claims about the heavens could be made, it was the privilege of the philosophers to make them; astronomers were excluded from this right. Astronomy, which held an inferior position in the hierarchy of disciplines compared to theology and philosophy (the latter was dominated by Aristotelian cosmology and Aristotelian physics), was associated with mathematics or mere numerical calculations. Its task was to produce quantitative predictions of planetary motions and to give mathematical explanations which would 'save the phenomena', i.e., which would bring the irregular motions observed in the heavens in agreement with a combination of uniform circular motions considered to be the real motions or forms as they had been ascertained by reason and thought.

In his endeavor to change this situation, Galileo had to tumble the old dichotomy between a celestial and a terrestrial sphere and turn the metaphysical universe reserved

---

2 Robert Westman discusses the role of astronomy in the social hierarchy of disciplines in the sixteenth and early seventeenth century in his article "The Astronomer's Role in the Sixteenth Century: A Preliminary Study," in *History of Science* 18 (1980), 105-47. See also his article "The Copernicans and the Churches," in David C. Lindberg and Ronald L. Numbers, eds., *God and Nature* (Berkeley: University of California Press, 1986), 76-113. For a discussion of Galileo's strategy to free himself from the restrictions imposed on astronomy, see Mario Biagioli, *Galileo Courtier: The Practice of Science in The Culture of Absolutism*, (Chicago: University of Chicago Press, 1993). Jean Dietz Moss observes that "[t]he elevation of the astronomer almost to divinity is a theme that Galileo was to carry to a dangerous point in the *Dialogue*." (*Novelties,* 50)

3 "La nuova scienza e la visione degli oggetti," in *Lettere italiane* 21, no. 3 (1969), 267, 269 and 280.

4 In the first place and traditionally, by the Catholic Church (which is the one that concerns us in Galileo's case), but, as Robert Westman shows, also by the protestant churches (though based on different biblical exegeses), and especially by Martin Luther and Philipp Melanchton, Luther's close friend and educational adviser.

to the theologians and the Aristotelian philosophers into a physical one beyond natural human reach and perception, but not beyond the reach of human comprehension.[5] He had to argue that the whole universe was essentially one in kind and was bound to basically the same laws in all of its parts. This would allow him to use the Earth as a cosmological paradigm and reference point for physical extrapolation. From there, he needed to develop a new kind of physics, which would account for terrestrial as well as celestial phenomena by applying the astronomer's mathematics to the sensual world.

Obviously, this project was way beyond the scope of the *Sidereus nuncius* and would not be fulfilled in Galileo's lifetime, nor in that of many of his successors. In fact, it was not until Newton's *Principia* that the development of a new physics would start to take its due course. Yet the *Sidereus nuncius* marked an important beginning and, like all beginnings, it needed to be convincing.

Apart from the great discoveries it relates, the *Sidereus nuncius* is remarkable for the clear and elegant style of its argument and the powerful rhetoric of its imagery. This is particularly true with respect to the first part of the book dedicated to the Moon, which will be at the center of our attention. In the following discussion, I will be focusing primarily on Galileo's use of imagery. I believe that his imagery is fundamental to the argument of the *Sidereus nuncius* not only for the way it illustrates Galileo's descriptions, but, more substantially, for fleshing out his argument in a very literal sense of this metaphor. It is to a large extent through his use of imagery, rather than argumentatively, that Galileo counters the dogmatic beliefs of the Aristotelian philosophers and tries to make his theory palatable to his audience.

In analyzing images seen through the telescope, Galileo regularly draws analogies to what he claims to be similar phenomena on Earth and then describes these phenomena in vivid images. At other times, he makes use of images to illustrate rather than to explain what he saw through the telescope. Such images are used exclusively for their pictorial value rather than for any intrinsic similarity between the telescopic and the descriptive image. Finally, there are images that show neither an intrinsic nor a pictorial relation to what Galileo saw, but which are only used for their affective, poetic power. The use of imagery in the *Sidereus nuncius* works toward three basic ends: (1) to create a physical world out of a visual image; (2) to establish an essential similarity between the Moon and the Earth; (3) and to translate the meaningless telescopic images into familiar images that bear significance in the reader's mind. In so doing, it addresses some peculiar epistemological problems that the telescope and the kind of visual evidence it furnished posed: for one thing, the evidence was exclusively visual; for another, it was not evident, at least not self-evident.

In its complete restriction to visual evidence, the discovery of the heavens differed substantially from the many discoveries of previously unknown parts of the Earth made in this period. While these new territories remained unknown until the moment of their discovery, but then became immediately accessible, the Moon, the planets and the stars had been challenging human curiosity with their visual presence for thousands of years, but they remained inaccessible. The telescope, as revolutionary an invention as it

---

5 Fernand Hallyn suggests that the erasure of the opposition between the sublunary and the supralunary world turned a metonymic conception of space into one characterized by synecdoche. (*Poetic Structure*, 105-25)

was, did not actually change anything about this. In fact, it only made the restriction more obvious. For while travelers could explore distant lands in person and could add samples of plants or animals, even human beings, as proof and illustration of their accounts, astronomers using telescopes to explore the sky had to rely on their eyes only and they could never see enough. Their journeys remained metaphorical.[6]

The question Galileo faces is how best to convince his audience that his interpretation of what he saw through the telescope was correct. The rhetoric of his lunar descriptions reveals an attempt to overcome the great distance between the Earth and the celestial bodies, and to turn the visual closeness produced by the telescope into a physical one.

Another problem was that the telescope did not show mountains and valleys on the Moon, but offered a visual experience quite different from anything the human eye was accustomed to.[7] This made it more difficult to convince the public of a substantial similarity between Earth and Moon. Galileo therefore had to show that what *looked* strange through the telescope was actually not so. He needed to translate the telescopic image into familiar images of terrestrial phenomena not only by drawing analogies to

---

6 The teasing inaccessibility of the stars became the theme of several poetic works that were inspired by the new astronomical discoveries. One of the most striking instances I have come across is in "Cupid's Conflict," a minor poem by Henry More, where Cupid is described as playfully touching and embracing the stars: "My mighty wings high stretch'd then clapping light,/I brush the starres and make them shine more bright./Then all the works of God with close embrace/I dearly hug in my enlargèd arms." Cf. *Philosophical Poems of Henry More*, ed. Geoffrey Bullough (Manchester: Manchester University Press, 1931), 113. In a similarly playful way, Kepler asks in the Preface to his *Dioptrice*, the treatise published shortly after the *Sidereus nuncius* in which he explains how the telescope functions in theory: "What now, dear reader, shall we make out of our telescope? Shall we make a Mercury's magic-wand to cross the liquid ether with, and, like Lucian, lead a colony to the uninhabited evening star, allured by the sweetness of the place?" (Quoted by Nicolson in *Science and Imagination,* 40)

7 Various attempts have been made to explain what induced Galileo to interpret the telescopic images the way he did. Samuel J. Edgerton suggests that it was Galileo's background in drawing and perspective that prompted him to see the dark and light patches on the lunar surface as patterns of reflected light and shadow. Cf. "Galileo, Florentine 'Disegno', and the 'Strange Spottednesse' of the Moon," in *Art Bulletin* 44 (1984): 225-32, and *idem*, *The Heritage of Giotto's Geometry* (Ithaca: Cornell University Press, 1991), 223-53. Others have argued that Galileo was already a Copernican when he started out with his telescopic observations and that it was this mind-set which made him recognize the four light spots in Jupiter's vicinity as satellites and which made him see the Moon as another Earth. See for example I. Bernard Cohen, "What Galileo Saw: The Experience of Looking through a Telescope," in *From Galileo's 'Occhialino' to Optoelectronics* (International Conference at the University of Padova, 1992), 445-72.

The English scientist Thomas Harriot made a sketch of his lunar observations with a telescope that predated the publication of the *Sidereus nuncius*, yet the sketch was never published. As Terrie F. Bloom has pointed out, Harriot's drawing of the lunar surface shows that he had no clear understanding of the phenomena that he was transcribing onto paper and did not interpret the spots he saw as shadows. Bloom argues that the reception of Galileo's *Sidereus nuncius* played in fact an essential role in transforming Harriot's perception of the Moon's features, as can be witnessed from Harriot's second and later drawing of the moon. Cf. "Borrowed Perceptions: Harriot's Maps of the Moon," in *Journal for the History of Astronomy* 9 (1978): 117-22. Norriss Hetherington calls this "a particularly good example of the transformation of belief into seeing." Cf. *Science and Objectivity* (Ames: Iowa State University Press, 1988), 16.

similar phenomena on Earth, but by evoking their image in the imagination of his readers so forcefully as to make them exchange their original experience for familiar ones. He had to predispose his readers so that they would immediately see something familiar. The images had to be in the readers' mind before they actually looked through a telescope themselves.[8] They also needed to be vividly impressed in the minds of those readers who did not have access to a telescope.[9] To this end Galileo's rhetoric was indispensable.

A discussion and illustration of Galileo's rhetoric of imagery focusing on the issues I have just summarized will be offered in the first part of this chapter. I will also at times consider passages from the *Dialogo sopra i due massimi sistemi del mondo* (*Dialogue*), where Galileo presented his lunar theory in more extended form and set it more explicitly within the context of Copernican theory. However, we also need to consider that Galileo's discoveries and the possibilities the telescope offered had more unsettling implications. I have just mentioned the double-edged character of the telescope's achievement: On the one side, there was its ability to bring remote objects closer and make previously unperceived ones visible; on the other, the evidence it produced was exclusively visual in nature. Moreover, while the telescope did show more than was visible with the naked eye, it never showed enough and always implied more.[10] The telescope thus drew attention to its limitations while simultaneously suggesting that the potential for new discoveries may be inexhaustible. In so doing, it posed a threat to the geocentric cosmology in its Stoic and Christian adaptation, a threat that was a great deal more devastating than that posed by Copernican theory alone.

The Stoics, followed by the Christian exegetes, had used the Aristotelian geocentric cosmos as a metaphor for their anthropocentric teleology. According to both Stoic and Christian thought, man existed in a metaphysically distinguished place in the world and could rightfully relate the entire cosmic spectacle to himself. The cosmos had been created exclusively for and because of man, whose highest destination in life was that

---

8 It is, therefore, not surprising that Galileo preferred to be present when others tried to re-create his observations. See for example his letter to Belisario Vinta, March 13, 1610, in *Opere*, vol. 10, 289.

9 See Steven Shapin's discussion of the problem of replicating evidence and the attempt to bridge it through "virtual witnessing": This "involves the production in a reader's mind of such an image of an experimental scene as obviates the necessity for either its direct witness or its replication. Through virtual witnessing the multiplication of witnesses could be in principle unlimited." ("Pump and Circumstance," 491) An interesting discussion of the laborious process involved in crafting astronomical drawings in nineteenth-century nebular astronomy is offered by Simon Schaffer in "On Astronomical Drawing," in Caroline A. Jones and Peter Galison, eds., *Picturing Science, Producing Art*, (New York: Routledge, 1998), 441-74.

10 To some extent, both of these latter points are still valid today, yet they have been qualified by various factors: by the invention of radio telescopes, for instance, which respond to frequencies of electromagnetic radiation that our senses do not detect at all and provide an output from which we can derive information about objects that radiate at these frequencies. Cf. Harold I. Brown, "Galileo on the Telescope and the Eye," in *Journal of the History of Ideas* 46 (1985): 487. Then there are space telescopes which, even though still limited to a minuscule particle of the universe, allow at least some change of perspective. Finally, there is space travel and exploration by satellites, which, though again extremely limited in range, offer information in addition to that offered by telescopes.

of being the *contemplator caeli.*[11] This meant that the cosmos only acquired its full significance through man's contemplation and appreciation of it and would lack meaning as well as purpose without it. It could, therefore, not comprise parts that were beyond the reach of human perception. By showing as well as implying that it did and that some of its parts might even be forever beyond the reach of human perception, the telescope demonstrated that this idea was invalid. In conjunction with Copernican theory, it dismissed the possibility of relating the whole universe to man as the one and only center of creation.

I am not concerned here with the consequences this turn had for anthropocentric teleology and Christian theology – obviously, they were devastating. I am interested in the way in which Galileo's telescopic discoveries suggest the idea, the necessity even, of a change in perspective, of taking up another standpoint – physically as well as imaginatively: If there are objects in the universe that the human eye cannot reach from the Earth, it follows that more or other objects would be visible from a different standpoint. The idea of a change in perspective also implies the possibility of perceiving things differently, of different norms and orders, of things *being* different – not in a metaphysical sense as the Aristotelians asserted, but in a physical one. Moreover, it implied the possibility of the inadequacy of human imagination in drawing their image in the mind.

This sheds an awkward light on Galileo's use of imagery and his emphasis on sameness rather than difference in the *Sidereus nuncius.* Yet Galileo must have been the first to notice this since, as he observed on various occasions, the principles of nature's complexity and of the inexhaustibility of discoveries to be made about the universe were fundamental to his conception of natural philosophy. Indeed, we shall see not only how Galileo's use of imagery is frequently irreconcilable with the arguments put forth in the *Sidereus nuncius*, but how its adequacy is questioned or at least qualified in the context of the definition of the possibilities and limits of natural philosophy (understood here as the kind of knowledge that comes from studying the universe) Galileo gives elsewhere.

How much Galileo was actually aware of the possibility of difference in the universe is evident from a passage in the *Dialogue*, where he addresses not only the issue of diversity, but also that of the nature and limits of human imagination: When the discussion turns to whether the Moon houses any plants, animals or human beings similar to the ones existing on Earth, Sagredo (who poses as the *soi-disant* neutral, but insightful third party in the dialogue) feels inclined to a negative answer. But then he muses that things on the Moon might be so fundamentally different from what they are on Earth that they are entirely beyond the grasp of the imagination. He uses the figure of analogy to give an example of "a person born and raised in the forest among wild

---

[11] Hans Blumenberg, *Paradigmen zu einer Metaphorologie* (Bonn: Bouvier, 1960), chapter ix: "Metaphorisierte Kosmologie." Blumenberg argues that the shift from the geocentric to the heliocentric cosmology was understood as a metaphor, an absolute metaphor, to be used in response to the question as to man's position in the universe, his central predestination versus his role as mere hanger-on in the world mechanism, and thus the question regarding the relation between man and all remaining being.

beasts and birds, and knowing nothing of the watery element",[12] and argues that such a man would never, not even with the liveliest imagination (*perspicacissima immaginazione*), be able to imagine to himself the sea with all its creatures or palaces and cities built and inhabited by people like him. He concludes with the passage I quoted at the beginning of the Introduction:

> Thus, and more so, might it happen that in the Moon, separated from us by so much greater an interval and made of materials perhaps much different from those on Earth, substances exist and actions occur which are not merely remote from but completely beyond all our imaginings, lacking any resemblance to ours and therefore being entirely unthinkable. *For that which we imagine must be either something already seen or a composite of things and parts of things seen at different times; such are sphinxes, sirens, chimeras, centaurs, etc.*[13]

Salviati (who in the dialogue more or less represents Galileo himself) concludes that "it is indeed possible to discover some things that do not and cannot exist on the Moon, but none which I believe can be and are there, except very generally."[14] Then he goes on to list these general attributes.[15]

---

[12] "uno nato e nutrito in una selva immensa, tra fiere ed uccelli, e che non avesse cognizione alcuna dell'elemento dell'acqua". (*Opere*, vol. 7, 86.)

[13] My emphasis. This passage evokes the beginning of the *Ars poetica*, where Horace cautions against imaginative creations that violate decorum: "If a painter chose to join a human head to the neck of a horse, and to spread feathers of many a hue over limbs picked up now here now there, so that what at the top is a lovely woman ends below in a black and ugly fish, could you, my friends, if favoured with a private view, refrain from laughing? Believe me, dear Pisos, quite like such pictures would be a book, whose idle fancies shall be shaped like a sick man's dreams, so that neither head nor foot can be assigned to a single shape." Cf. Horace, *Satire, Epistles, Ars poetica*, trans. H. Rushton Fairclough (Cambridge: Harvard University Press, 1961), 451.

[14] "e finalmente mi pare di poter ritrovar bene alcune delle cose che non sieno nè possan esser nella Luna, ma non già veruna di quelle che io creda ch vi sieno e possano essere, se non con una larghissima generalità" (*ibid.*).

[15] *Dialogue Concerning the Two Chief World Systems*, trans. Stillman Drake (Berkeley: University of California Press, 1967), 61-62. Unless otherwise noted, English translations of the *Dialogue* are taken from this edition, henceforth cited as *D*.

The whole concept of the inaccessibility of human imagination to the unfamiliar including the particular example given by Sagredo originates from Plutarch's *De facie*, which Galileo had used as a source, with the difference that Sagredo and Salviati are much more careful in voicing any positive opinion regarding forms of life on the Moon similar to those on Earth. In fact, the speaker in Plutarch's dialogue turns the tables by saying that "[t]hose men [meaning the inhabitants of the Moon], I think, would be much more amazed at the Earth, when they look out at the sediment and dregs of the universe, as it were, obscurely visible in moisture, mists, and clouds as a lightless, low, and motionless spot, to think that it engenders and nourishes animate beings which partake of motion, breath, and warmth." (Plutarch, "Concerning the Face," 179) For a discussion of Galileo's adoption and development of various subjects already treated by Plutarch in the *Sidereus nuncius* and the *Dialogue*, see Paolo Casini, "Plutarco, Galileo e la faccia della luna," in *Intersezioni* 4, no. 2 (1984): 397-404. Casini argues that Galileo possessed a copy of Xylander's Latin translation of Plutarch's text, from which he at times quoted literally. Kepler owned the same translation of Plutarch (see his letter to Matthias Bernegger of 4 December 1623; *Gesammelte Werke,* ed. W. von Dyck and M. Caspar *et al.* [Munich: Beck, 1937-], vol. 18, 143). This may in part explain why he

Both in the *Dialogue* and in the *Sidereus nuncius*, Galileo carefully and diplomatically avoids any fantastical digressions into what life might look and be like up there and strictly limits himself to giving what he considers well-founded and legitimate interpretations of his observations through the telescope or with the naked eye. However, he is still faced with the question of how he should best translate these interpretations into words and images that will convince his audience of the correctness of his interpretations and enable them to visualize what he describes. As a consequence, the images Galileo evokes from his descriptions in illustration of the lunar world resemble to a considerable degree the kind of 'chimeras' Sagredo refers to in the *Dialogue*. They do so at least in that they are metaphorical in nature, i.e., representative, in the first place, of something else, and in that they are used to transgress the bounds of knowledge and imagination.

I will try to show how one may reconcile Galileo's implicit criticism of the anthropocentric character of rhetorical imagery with his conception of a natural philosophy that is concerned with the sensible world as opposed to a philosophy that only produces chimeras of the mind. However, Galileo's first concern in the *Sidereus nuncius* was to do away with the notion of a sharp dichotomy between a terrestrial and a celestial realm, and to establish a substantial relation between the Earth and the rest of the universe. It is clear that this is not easily combined with an argument for the possibility of difference. Before we therefore witness the deconstruction of Galileo's rhetoric, let us see how he constructs his argument against the Aristotelians in the first place.

Galileo describes his first discovery of the innumerable stars of the Milky Way as if he had literally bumped into them: "if you direct a glass to any of [the many weakly shining colored patches in the sky] you will meet with a dense crowd of stars."[16] The English translation of this passage only weakly conveys the sense of the original wording, "si in illarum quamlibet Specillum convertas, *Stellarum constipatarum coetum offendes*", in which we get a much stronger sense of an actual physical collision.[17] Shortly before, in describing the multitude of stars that became visible when he directed his telescope on the constellation of the Pleiades, Galileo exploits the optical illusion of closeness between the individual stars to give the impression of stars packed tightly together like sardines in a very confined space. Again, the idea of physical constriction is rendered most powerfully in the original wording: "sex Stellas Tauri, PLEIADAS dictas, dipinximus ... intra angustissimos in coelo cancellos obclusas, quibus

---

immediately recognized Galileo's source even though Galileo had not indicated it. Cf. *Kepler's Conversation with Galileo's Sidereal Messenger*, ed. and trans. Edward Rosen (New York: Johnson Reprint Corporation, 1965).

16 Galileo, *Sidereus nuncius*, trans. Albert Van Helden (Chicago: University of Chicago Press, 1989), 62. Explanation mine. Unless otherwise noted, English translations of the *Sidereus nuncius* are taken from this edition, henceforth cited as *SN*.

17 *Opere*, vol. 3, pt. 1, 78. Emphasis mine. Maria Timpanaro Cardini's Italian translation of this passage renders the original very well; she translates "se in una qualsiasi di esse si rivolga il cannocchiale, *c'imbatteremo in un fitto ammasso di stelle*." Cf. *Sidereus nuncius*, trans. Maria Timpanaro Cardini, ed. Andrea Battistini (Venice: Marsilio, 1993), 129. Emphasis mine.

aliae, plures quam quadraginta, invisibiles adiacent."[18] The language used to describe the visual image turns the heavens into a world of matter.

The notion of concreteness is already evoked in the book's title, which through the ambiguity of the word *'nuncius'* carries a double meaning. There has been a long-standing debate among critics as to whether one should read *'nuncius'* exclusively as 'message' or if the second meaning of the word, 'messenger', is also valid. The main contenders in the debate on the correct interpretation of the title are Edward Rosen and Stillman Drake. Did Galileo accidentally choose this ambiguous word (Rosen's position) or did he intend the ambiguity and only later, on being attacked for the presumption of fashioning himself as the messenger from the stars, meekly deny any such intention (Drake's position)?[19]

Without wanting to review too much of the debate, let me just briefly state here that I find Drake's argument more convincing since he takes into account the general predilection for puns and *double entendres* at the time, which Galileo, otherwise known for his love for precision in language, must have shared to some extent. Moreover, he points out that Galileo was perfectly aware of the fact that some readers, among them most notably Kepler, had taken the word *'nuncius'* to mean 'messenger', and that Galileo only objected to this interpretation when he was attacked for being presumptuous.[20] Mainly, however, I think that the second meaning of the title participates in the recurrent rhetoric of physical contact Galileo uses in describing the heavenly bodies he perceived through the telescope. This rhetoric emphasizes the fact that he is speaking about the Moon, the planets and stars as physical objects and not only as visual phenomena.

The idea of a material lunar world of the kind we are familiar with is conveyed most vividly and unambiguously in the descriptions of the Moon and in the analogies that are drawn between lunar and terrestrial phenomena. In the *Dialogue*, when after a clever demonstration made by Salviati to prove that in order to reflect the Sun's rays as brilliantly as it does, the Moon must have a rough surface, Sagredo enthusiastically exclaims: "If I were on the Moon itself I do not believe that I could touch the roughness of its surface with my hand more definitely than I now perceive it by understand-

---

18 *Opere*, vol. 3, pt. 1, 78.

19 Rosen, "The Title of Galileo's *Sidereus nuncius*," in *Isis* 41 (1950), 287-89; *idem*, "Review: Stillman Drake's *Discoveries and Opinions of Galileo*," in *Journal of the History of Ideas* 18 (1957): 439-48; as well as the introduction to Rosen's edition of Kepler's *Conversation*; Drake, "The Starry Messenger," in *Isis* 49, no. 3 (1958): 346-47; for a summary of both their arguments, see introduction to Isabelle Pantin's edition of the *Sidereus nuncius* (Galileo Galilei, *Le méssager céleste* [Paris: Belles Lettres, 1992], XXXIII-XXXV). See also François Russo, "Note sur la traduction du titre de l'ouvrage de Galilée, *Sidereus nuncius*," in *Revue d'histoire des sciences* 20 (1967): 67-69. Galileo was primarily attacked by Orazio Grassi, Martin Horky and Francesco Sizzi (cf. Pantin's edition of *SN*, XXXV).

20 For this, see also Albert Van Helden's introduction to *SN*, 20. In a study of Galileo's emblematic program in the *Sidereus nuncius*, Mario Biagioli argues that in order to dramatize his presentation of the 'Medicean stars' as an encounter between the Medici and their destiny rather than his own discovery, Galileo must be given the status of starry ambassador; only then does his discovery become an omen from the stars (a *sidereus nuncius*). Cf. "Galileo the Emblem Maker," in *Isis* 81 (1990): 245.

ing your argument."[21] Yet the validity of such analogies depends on the existence of a similarity in substance between the Earth and the Moon, which first needed to be established. Since Galileo had no means to prove it except by such analogies and since the Aristotelians would dismiss this kind of evidence with the dogmatic principle of a substantial *difference* between Earth and Moon, there was no way that he could establish this similarity argumentatively. He therefore established it rhetorically by constructing a relation of kinship between Earth and Moon.

In a first step toward constructing a relation of kinship, Galileo personifies the Moon and the Earth, thus bringing them to life. With regard to the Moon, the most recurrent instance of this personification can be found in the reference to an image that has become such a cliché that it is easily overlooked: the *face* of the Moon (*facies Lunae*). It is quite remarkable that Galileo should choose to use this folkloristic image at all as it stands out oddly against his commitment in the *Sidereus nuncius* to do away with all the myths surrounding the Moon – one of the most common of which was that the Moon had a face and that it housed the 'Man in the Moon' – and to describe the lunar surface realistically.[22]

The reference to the Moon's face is only made indirectly as the Latin word *facies* is ambiguous: Besides meaning 'face', it also carries the more general meanings 'shape', 'form', 'appearance'. While the last meaning, especially, is also present in Galileo's use of the word, he clearly exploits the ambiguity: His reference to the *facies Lunae* is, first of all, an implicit reference to Plutarch's *De facie in orbe lunae*, a dialogue on different theories regarding the nature of the Moon, where one of the interlocutors describes the lunar surface as rugged and argues that it reflects solar light. In the title of Plutarch's text, the meaning of the word *facies* is limited to 'face'. Galileo never explicitly refers to this text, but he uses many issues raised therein as a kind of blueprint for the *Sidereus nuncius* and at times even makes literal quotations from it. Indeed, to an informed reader like Kepler, the parallelisms were obvious.[23]

Yet there is more to Galileo's use of this image, as becomes clear when we observe how the image is elaborated in the successive descriptions of the way light and dark spots keep changing on the lunar surface. In these descriptions we witness how in this "bright face", "uneven, rough, and crowded with depressions and bulges", "bright points light up, as if they are sprouting ... grow", "bright excrescences" extend themselves and "a great number of small darkish spots ... are distributed everywhere." Galileo brings the dead metaphor to life, turning the unchanging, mystical face into a site of wild spectacle.[24]

---

21 *D*, 73. – "Se io fussi nella Luna stessa, non credo che io potessi con mano toccar più chiaramente l'asprezza della sua superficie di quel ch'io me la scorga ora con l'apprensione del discorso." The phrase 'ch'io *me la* scorgo' in the Italian conveys a special sense of familiarity. (*Opere*, vol. 7, 98, emphasis mine)

22 With respect to these myths, Salviati says disparagingly: "Of the Moon's appearance, I find no more exact description than that some say it represents a human face; others, that it is like the muzzle of a lion; still others, that it is Cain with a bundle of thorns on his back." (*D*, 49)

23 Cf. note 15, *supra*.

24 *SN*, 39-43. – "luminosa facie[s]", "inaequalem, asperam, cavitatibus tumoribusque confertam" "excrescentiae lucidae" ... "interim tamen aliae atque aliae [i.e., "lucidae cuspides"], hinc inde quasi pullulantes, intra tenebrosam partem accenduntur, augentur, ac demum eidem luminosae

The metaphorical image of the lively face is here masterfully combined with a complex and clever argument concerning celestial immutability and the question of lunar substance: We have to bear in mind that what Galileo saw through the telescope were not cracks and excrescences as such, but changing patterns of light and dark spots on the lunar surface. Moreover, the changes Galileo observed took place much more slowly and were therefore never quite as dramatic in effect. By creating a quick-motion effect of the kind a modern time-lapse camera would produce, Galileo emphasizes how conditions of illumination are in constant change on the lunar surface, how dark and light patches leak into each other and how in a short period of time a light spot may turn dark and vice versa. As we know, the explanation Galileo provides for this phenomenon is that the surface of the Moon is not even, but rugged, so that the light is not reflected evenly off the surface.

Yet the Aristotelians did not only claim that the lunar globe was perfectly spherical, they also rejected the thesis that its surface reflected sunlight. The cleverness of Galileo's line of argumentation lies therefore not so much in the explanation itself, for it does not defeat the lunar theories of the Aristotelians: Galileo's explanation could and would be dismissed by a counter-explanation which accounts for the existence of the darker and lighter spots. One such explanation held that different parts of the lunar surface receive the sunlight differently in proportion to their density and emit light by becoming actively luminous.[25] The cleverness of Galileo's argument lies in its emphasis on change: Galileo's explanation for the phenomenon could not be dismissed without the Aristotelians at the same time having to admit that, provided their theory of density is true, the density of the Moon would have to change constantly in order to account for the appearance and disappearance of light and dark spots. Accordingly, Galileo forces his opponents either to grant that the lunar surface is rugged and reflects solar light in the way the Earth does, or to admit that it undergoes changes, a point that obviously contradicts the principle of celestial immutability. Unable to offer irrefutable proof for his own theory, he makes an attempt to cause havoc in the dogmatic principles of his opponents.[26]

---

superficiei, magis adhuc extensae, copulantur." ... "magna nigricantium macularum exiguarum copia, ... undiquaque conspergit". (*Opere*, vol. 7, 62-64, explanation and adaptation mine)

In his "Geographical or, if You Prefer, Selenographical Appendix" to the *Somnium*, Kepler adopts the image of the face and puts it in more glaring terms: "the spots located near the line of separation look to the eye not unlike a boy's face, disfigured by pimples bulging far out, if this swollen face is illuminated by light whose rays come from either side." Cf. *Kepler's Somnium or Posthumous Work on Lunar Astronomy*, ed. and trans. Edward Rosen (Madison: University of Wisconsin Press, 1967), 160.

25 This latter view was proposed among others by the Aristotelian professor Fortunio Liceti in a book on light of 1640. Liceti based his theory on the phenomenon of luminescent rocks displayed in the year 1611 in Rome by Galileo himself. The rocks had caused a great stir among intellectuals since it demonstrated "that light was a phenomenon separable from the idea of heat and from the presence of a luminous environment." Cf. Pietro Redondi, *Galileo Heretic*, trans. Raymond Rosenthal (Princeton, NJ: Princeton University Press, 1987), 9. For Liceti and his theory of light, see pp. 20-27.

26 In his article "Galileo's Lunar Observations in the Context of Medieval Lunar Theory," Roger Ariew offers an acute summary of the Scholastic position of Galileo's opponents, which, as he shows, was largely influenced by Averroes' interpretation of *De caelo*. Based on this summary,

Yet what is the use of the facial image in all this? By describing the rugged surface of the Moon in the figure of a face, Galileo in the first place points out how obvious his theory is: rather than making up some obscure theory about the density of lunar substance or other hidden characteristics of its body, he is only describing what is apparent, i.e., the visible features of a face that cannot be overlooked. Here Galileo is really beating the Aristotelians at their own game, for unlike them, he is not making claims about the essential nature of lunar substance, but is only describing the Moon's appearance and pointing out similarities to phenomena that may be observed on Earth. Secondly, and here argument is entirely abandoned for rhetoric, the image of the face is used to establish the similarity between Earth and Moon and should be read in the context of the metaphor of kinship between them which is used at various points in the text: It is in the face that we recognize the kind of similarities attributable to kinship, and the characteristics of the lunar face clearly manifest the Moon's kinship to the Earth.

A good example of Galileo's use of the kinship metaphor is when he refers to the "cognatio atque similitudo inter Lunam atque Tellurem", '*cognatio*' meaning 'blood-relationship, consanguinity, kinship' or that by which a relationship is recognized.[27] The use of the metaphor reaches its climax when, in the explanation of the Moon's secondary light, the relation between Moon and Earth is described as a close and caring friendship, in which the two support each other with light as if to take shelter from the 'horror of the night': "In an equal and grateful exchange the Earth pays back the Moon with light equal to that which she receives from the Moon almost all the time in the deepest darkness of the night."[28] The explanation of the Moon's secondary light

---

Ariew argues that to a Scholastic Galileo's argument must have sounded curiously circular since the thesis that the Moon reflects light and is Earth-like in nature is argued for on the basis of the premise that its surface is rugged, while the ruggedness of the surface depends on the premise that the lunar light is caused by reflection. He concludes that, "[b]y themselves, Galileo's lunar observations were therefore not sufficient to refute Scholastic science, or to demolish the foundations of Aristotelian cosmology by destroying the distinction between the sublunar and celestial substances, as is often claimed." (225-26) Although I largely agree with this position, I think that Galileo's strongest weapon against the Aristotelian position was the observation that the spots keep changing, a point Ariew disregards. Cf. *Studies in the History and Philosophy of Science* 15, no. 3 (1984): 213-26.

Even more than in this debate around lunar spots, it was in the argument around the nature of sunspots which arose a little later between Galileo and the conservative Jesuit Christoph Scheiner that Galileo upheld a dangerous position: He argued that these spots were on or near the surface of the Sun. This implied that the purest celestial body, and with it the heavens, could not be perfect. Cf. Albert Van Helden, "Galileo and Scheiner on Sunspots: A Case Study in the Visual Language of Astronomy," in *Proceedings of the American Philosophical Society* 140, no. 3 (1995): 358-96. For the argument that in some theological circles Galileo's lunar discoveries were actually welcomed as corroboration of an alternative exegetical tradition of the symbolic significance of the Moon, see Ostrow, "Cigoli's *Immacolata*," and *idem*, *Art and Spirituality*, 243.

27 *Opere*, vol. 3, pt. 1, 72. For this translation of '*cognatio*', cf. *Oxford Latin Dictionary*, ed. G. W. Glare (Oxford: Clarendon Press, 1982). In his *Conversation*, Kepler also uses the phrase "kinship of the Moon with the Earth," arguing that already his teacher Michael Mästlin had pointed out such a relationship. (30)

28 *SN*, 55. – "aequa grataque permutatione rependit Tellus parem illuminationem ipsi Lunae, qualem et ipsa a Luna in profundioribus noctis tenebris toto fere tempore recipit." (*Opere*, vol. 3, pt. 1, 74)

is taken up again and elaborated by Salviati and Sagredo in the *Dialogue* with the effect that Simplicio, the Aristotelian and uncomfortable third party among the disputants, expresses his reluctance toward "this companionship between the Earth and Moon of which you want to persuade me."[29]

Earlier in the *Dialogue*, when the discussion turns onto the question whether generations and alterations only take place on Earth while the heavens are immutable, Sagredo rejects this idea and comments: "Otherwise I do not see how the influence of the Moon or the Sun in causing generations on the Earth would differ from placing a marble statue besides a woman and expecting children from such a union."[30] The personification of Earth, Moon and Sun has thus reached the stage where it has turned into both a sexual and fruitful relationship between man and woman. Timothy Reiss has pointed out that Galileo's discoveries are repeatedly described in terms of a fight between nature and the scientist who tries to get at her secrets. In the case of the Moon, it is a fight between the male astronomer and the female Moon, in which the astronomer lays bare the woman's most secret parts. Both in the *Sidereus nuncius* and in the *Dialogue*, the Moon is a woman that must be raped by the astronomer's penetrating gaze.[31]

Through the metaphor of kinship Galileo creates the impression that the world out there is not as remote and strange as traditional theories held. Within the framework of Copernican theory, it has become possible to conceive the relationship between Earth and Moon in terms of identity rather than alterity, and Galileo proposes that we visualize the lunar world with familiar terrestrial features.[32] This proposition is made most plainly when the lunar surface is described in terms denoting geographical phenomena

---

A little earlier, we find the expression *nocturnus horror* (72). For a discussion of the terror felt by some at the thought of 'the terror and silence of infinite spaces', see Nicolson, *Science and Imagination*, 30-57. The phrase 'nocturnus horror' may also allude to the phrase in the *Vulgate* "non timebis a timore nocturno" (Psalm 90:5).

29 *SN*, 96. – "questa società che voi vorreste persuadermi tra la Terra e la Luna". (*Opere*, vol. 3, pt. 1, 122)

30 *D*, 60. – "altramente non so intendere che l'applicazione della Luna o del Sole alla Terra per far le generazioni fusse altro che mettere a canto alla sposa una statua di marmo, e da tal congiugnimento stare attendendo prole." (*Opere*, vol. 7, 84-85)

31 "Espaces," 15. Paolo Rossi argues that in the early modern era "there was a continuous discussion, with an insistence that bordered on monotony, about a logic of [discovery] conceived as a *venatio*, a hunt – as an attempt to penetrate territories never known or explored before." Cf. *Philosophy, Technology, and the Arts in the Early Modern Era*, trans. Salvator Attanasio (New York: Harper and Row, 1970), 70. See also William Eamon, who argues that this "transformation of the logic of scientific discovery ... resulted from a shift in the institutional locus of scientific activity from the universities ... to places intertwined with networks of personal patronage." Court scientists practiced a style that glorified the reputation of their prince. This effected a revalidation of *curiositas* from vice into virtue. Cf. "Court, Academy, and Printing House: Patronage and Scientific Careers in Late Renaissance Italy," in Bruce Moran, ed., *Patronage and Institutions* (Woodbridge: Boydell Press, 1991), 27.

32 Enthusiastic proponents of the theory of the plurality of worlds later used Galileo's analogy between Earth and Moon as only the first step in an argument that they extended to other planets and solar systems. Cf. Gianguido Piazza, "La Luna della Luna: Metafore, congetture e scoperte della moderna astronomia planetaria," in Nadia Minerva, ed., *La Luna allo specchio* (Bologna: Pàtron, 1990), 16, and chapter 5, *infra*.

on Earth such as 'mountains', 'plains' and 'seas'.[33] The language Galileo introduces is profoundly metaphorical in that the vocabulary generally adopted to represent one domain of reality is transferred onto another, up to that point felt as profoundly different. While all of these terms are used up to this day in the language of selenography and some of them literally, others such as *maria* have remained metaphors.

In his attempt to turn the Moon into a familiar terrestrial world, Galileo regularly explains phenomena visible on the lunar surface by drawing analogies to phenomena that we can witness on Earth with our own eyes. Thus, he explains the dark patches visible in the light part of the Moon close to the terminator (the dividing line between the illuminated and the dark section of the Moon) by drawing the comparison to deep valleys on Earth which the sunlight only falls upon once it has reached a steeper angle to the surface of the Earth. Correspondingly, he compares the light patches shining forth from the dark section near the terminator to the peaks of high mountains which catch the sunlight first, even though they stand beyond the range of the illuminated hemisphere.[34] By applying the Pythagorean theorem, Galileo even calculates the approximate height of these mountains.

While these analogies serve to contextualize visual impressions that bear direct witness to the thesis of the ruggedness of the lunar surface, there are other instances where analogies to terrestrial phenomena are used to explain why the telescopic image in fact does *not* bear signs of ruggedness and thus seemingly contradicts the thesis. Of this kind are the analogies drawn in explanation of how, even through the telescope, the circumference of the illuminated part of the Moon manifests a perfect sphericity instead of appearing like a "toothed wheel" (*sub specie quasi dentatae rotae*)[35]. Galileo offers two different explanations for this: he first explains it by drawing a comparison to many subsequent rows of mountains or tall waves in the ocean, which, if the observer is far away and situated at the same altitude, fuse together to a straight line on the horizon. His other explanation is based on the assumption that, just like the Earth, the Moon has an atmosphere of denser substance than the rest of the ether. Since on the circumference of the Moon this atmosphere is cut obliquely by our visual rays, this has the effect of blurring our perception of the Moon's ruggedness.[36]

Both explanations are set forth with admirable clarity and precision. This as well as the fact that they are accompanied by pictures evoking familiar visual experiences made on Earth is what makes them so convincing. Galileo claims that the corresponding experiences on Earth are of the same nature, only that in the latter case the image has been proven to be an optical illusion. However, this does not alter the fact that they are mere projections and, whether accurate or not, are not based on any positive visual evidence (for the circumference of the Moon appears smooth), but on the premise that the lunar surface is rugged. In turn, this premise serves as the point of departure for the

[33] The term 'seas' (*maria*) was introduced by Kepler in his *Conversation*, 27 and 106, note 224. In the *Dialogue*, Galileo toned down the analogies between Earth and Moon, pointing out the differences, or negative analogies, between the two rather than the similarities. Thus, Salviati notes the absence of clouds around the Moon, from which he concludes that it does not rain on the Moon. Galileo was generally inclined to think that there is no water on the Moon. (*D*, 99-100)

[34] *SN*, 41-42.

[35] *SN*, 53, and *Opere*, vol. 3, pt. 1, 69.

[36] *SN*, 48-51.

explanation why it does not appear so. Projections of this kind reveal a tendency to describe the Moon as more similar to the Earth than its visual image suggests.

There is even evidence that Galileo forged the resemblance between Earth and Moon and made the Moon more similar to the Earth than he had reason to believe it was. This leads us into a long-standing debate about the big circular crater that is so conspicuous in the two illustrative etchings of the Moon shortly before the first and the second quadrature (*fig. 1*) and the possible reasons behind it. I would like to suggest that it is Galileo's endeavor to describe the Moon as much as possible as another Earth which may account for this phenomenon. This argument requires some introduction.

The crater on the engraving is not only vastly exaggerated in size, but it is also singled out from a great number of craters visible in this area of the lunar surface. With regard to the exaggeration in size, Owen Gingerich argues that Galileo used this crater as a symbolic illustration for his argument rather than as a realistic representation of an actual crater. This seems convincing. Moreover, Gingerich suggests that because of the poor quality of his telescope, Galileo may only have been able to see this one crater, which left an indelible impression on him, but not the numerous other craters in the lower part of the lunar hemisphere.[37] This last part of Gingerich's argument does not convince me: looking at the area in which Galileo's big crater is located on a modern photograph of the lunar surface, one finds several other craters of similar or even bigger size[38] alongside it so that it seems unlikely that Galileo would have only been able to see this single crater, and that this crater would have been so conspicuous.

There are two indications that Galileo must in fact have seen a great number of craters in the lower lunar hemisphere. The first indication can be found in a passage that accompanies the engraving of the new Moon. The picture is very indistinct and does little justice to the precise verbal description that accompanies it.[39] Yet in the text Galileo mentions "a great abundance of dark spots in both horns, *especially in the lower one*"; and he goes on to describe with great precision the play of light and shadow on the circular craters:

> the dark part of the spot always faces the direction of the Sun and the *brighter border* surrounds the dark spot on the side turned away from the Sun and facing the dark part of the Moon.[40]

---

37 "Dissertatio cum Professore Righini et Sidereo Nuncio," in M. L. Righini Bosinelli and William R. Shea, eds., *Reason, Experiment, and Mysticism in the Scientific Revolution* (New York: Science History Publications, 1975), 77-88. Other articles in which the feature is discussed are Guglielmo Righini, "New Light on Galileo's Lunar Observation," *ibid.*, 59-76, to which Gingerich's article is a critical reaction, as well as Samuel J. Edgerton, "Galileo, Florentine 'Disegno'" and *idem*, *Giotto's Geometry*.

38 I am proceeding on the assumption that the crater depicted in the engraving is the Albategnius crater. For this view see Gingerich, "Dissertatio," as well as Ewen A. Whitaker, "Selenography in the Seventeenth Century," in René Taton, ed., *The General History of Astronomy* (Cambridge: Cambridge University Press, 1989), vol. 2, 124.

39 For Galileo's valuing the written word over pictures as a means of communication, see Mary G. Winkler and Albert Van Helden, "Representing the Heavens: Galileo and Visual Astronomy," in *Isis* 83 (1992): 195-217.

40 *SN*, 43. Emphases mine.

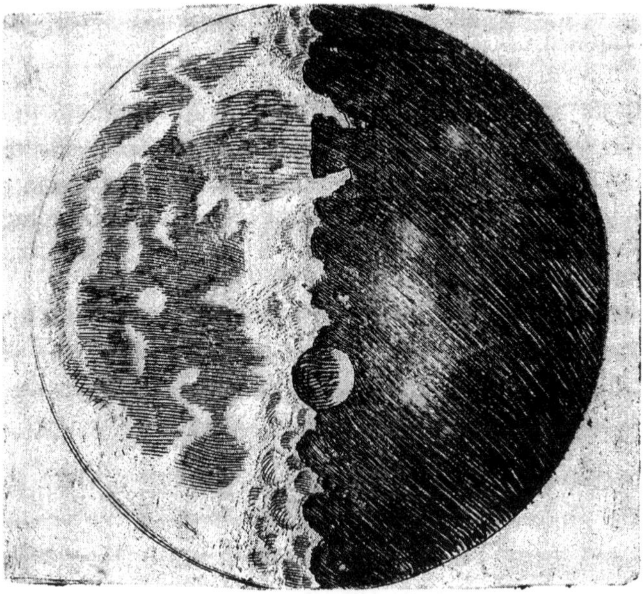

figure 1

figure 2

> Eratque magna obscurarum macularum vis in utroque cornu, maxime autem in inferiori: Semper tamen ... nigricans ipsius maculae pars irradiationis solaris locum respicit, splendidior vero limbus nigricantem maculam in parte Soli aversa, et Lunae tenebrosam plagam respiciente, circumdat.[41]

If Galileo saw so many craters so clearly on the terminator during new Moon, it is hardly credible that he only saw the single one shown on the engraving near first and second quarter and referred to in the text, as the craters on the terminator at first and second quarter are much more conspicuous than the ones on the terminator at new Moon. Yet he mentions no other craters in his description.

That Galileo did in fact see a great number of craters in this area of the lunar hemisphere becomes evident in one of Galileo's original sepia drawings which also shows the Moon near second quarter, the drawing he marked with the number "4" (*fig. 2*). In this drawing there are not only more craters, but several big ones aligned on the terminator, all of them showing the characteristic play of light and shadow on the circular cavity;[42] none of them is larger than the others to any particular degree. Most critics agree that Galileo made the sepia drawings shortly after he had made his telescopic observations, and they are generally considered to be much more accurate than the engravings included in the book. The latter were done by a professional engraver and have often been criticized for their lack of accuracy.[43] Therefore, if Galileo had noticed any crater of outstanding size during his telescopic observations, he surely would have captured it in his own drawings. The fact that he did not, shows that the great single crater was an afterthought.

As to the criticism that is so often voiced with respect to the inaccuracy of the engravings, it is important to note that the upper half of the engraving (reproduced in *fig. 1*) is a remarkably accurate copy of Galileo's original drawing. It is only the lower half which notably deviates from the original: Besides the exaggerated central crater, the area around the lower terminator has been given a scalloped appearance.[44] This not only suggests that Galileo's drawings served the engraver as copies, but that the deviations from the drawings in the engraving cannot be attributed to anything like the engraver's inaccuracy or lack of skill. It rather seems to indicate that Galileo had a hand in the matter.

Yet what could be his motive for changing the face of the Moon to such a considerable degree? Why should he wipe out the many other craters clearly depicted in his original drawing in favor of a single crater of enormously exaggerated size? Is the

---

41 *Opere*, vol. 3, pt. 1, 65.

42 While on the terminator in the engravings of the second quarter there are also great roundish irregularities that suggest a mountainous region, it is only in the big circular cavity that this characteristic play of light and shadow is retained. This implies that only this topographical feature has the form of a crater.

43 For this see Owen Gingerich, "Dissertatio," and Gingerich and Ewan A. Whitaker, "Galileo's Lunar Observations and the Dating of the Composition of the *Sidereus nuncius*," in *Journal of the History of Astronomy* 9 (1978): 155-69, as well as Samuel Edgerton, "Galileo, Florentine 'Disegno'" and *idem*, *Giotto's Geometry*, among others. For sharp criticism regarding the quality of the engravings see J. Classen "The First Map of the Moon," in *Sky and Telescope* 37 (1969): 82-84, and Zdenek Kopal, *The Moon* (Dordrecht: Reidel, 1969), 225.

44 Scott L. Montgomery, *The Scientific Voice* (New York: Guildford Press, 1996), 225.

explanation that he intended the big crater on the picture as a symbolic illustration for one of the several big lunar craters on the terminator really satisfactory? The fact that the crater is also singled out in the text while no other craters are mentioned seems to refute this.

There is a point that is easily overlooked in the whole debate, and it may at first seem either irrelevant or sound like a joke: Galileo never uses the term 'crater,' neither with respect to the circular 'spots' nor the big cavity, but keeps referring to 'valleys' and 'very high mountains'. It is true that because of their size and their large number these features resembled no terrestrial phenomena Galileo would ever have seen.[45] Galileo could, therefore, not have *recognized* them as craters. What is remarkable, however, is that he does not single them out for their peculiar, even enigmatic nature, that is, as being different from anything he knows. That would not fit his program, i.e., to emphasize *cognatio atque similitudo inter Lunam atque Tellurem*. Instead, he singles out one crater, magnifies it and compares it to the unique – but familiar – topography of Bohemia, thereby emphasizing that what looks strange on the Moon also exists on Earth. He in fact not only argues that the crater *resembles* the topography of Bohemia, but even claims that this is the way the region of Bohemia would look from outer space. (47) The crater therefore *is* Bohemia.

Apparently, Galileo had never even been to Bohemia himself, but possessed a copy of Ptolemy's *Geographia* in which it was described in the following terms: "the Black Forest surrounds the whole of Bohemia, enclosing it all round, like an amphitheater."[46] An Italian edition of Ptolemy's *Geographia* widely distributed in Padova at the time includes a map that depicts Bohemia as a perfectly round plane almost entirely fenced in by a unique circle of mountains.[47] It is now less difficult to see why Galileo had enlarged the crater and reduced the larger number of craters to a single one: a feature of mountains laid out in a circle such as Bohemia must be a rare phenomenon, not likely to be met with several times in close proximity. To claim that on the Moon it could be found by the hundreds,[48] would have sounded little convincing.[49] Galileo therefore had to devise a map of the Moon in which only one such feature appears.[50]

---

45 The craters on the Moon that are visible through a telescope are meteorite craters and not volcanic craters. Many of them are of enormous size due to the large size and the great impact of the object that hit the lunar surface. They date from a period of heavy bombardment which took place between 3.9 and 4.4 billion years ago and which must have affected the Earth just as much. Yet while on the Moon these craters have preserved their form (with the exception of those that were demolished by new impacts), on Earth they have almost entirely disappeared due to atmospheric influences. Cf. Michael Zeilik and John Gaustad, eds., *Astronomy: The Cosmic Perspective* (New York: Harper & Row, 1983), 551-52.

46 Galileo Galilei, *Le méssager céleste*, 66, note 49. Translation from the French mine.

47 Johannes Kepler, *Discussione col Nuncio Sidereo*, ed. Elio Pasoli and Giorgio Tabarroni (Torino: Bottega d'Erasmo, 1972), 131, note 106.

48 I am speaking in terms of what Galileo's still primitive telescope enabled him to see. Today we know that the number of craters on the Moon is much vaster.

49 This is why I only partly agree with Samuel Edgerton when he argues that the crater's exaggerated size in the engravings shows that Galileo permitted the engraver "a certain artistic license to exaggerate the size of ... [the] crater," which would allow him "to compare its steep sides with those of the high mountains surrounding the region of Bohemia. Thus he bade his artist to render it large, to dramatize that the moon is covered all over with such rugged depressions." (*Giotto's*

There is, I believe, another reason why Galileo chose the analogy with the region of Bohemia. While Galileo had no first-hand knowledge of Bohemia and the same is likely to be true for the majority of his readers at the time, the feature was a well established fact in *the* classical work of geography, Ptolemy's *Geographia*. The more learned among Galileo's readers could therefore be expected to be familiar with the feature through this source. In this way, Galileo's argument partakes of the authority of a classic. It is as if the fact that Ptolemy's name vouched for the correctness of the description legitimized the analogy itself.[51] Galileo's reference to Ptolemy may represent an attempt to benefit from his authority so as to forestall objections about his undermining past knowledge. It is, of course, highly ironic that Galileo should be relying on the authority of the man whose cosmology he was determined to overthrow. At the same time, it is largely in keeping with the provenance of other images of his.

Despite Galileo's artistry, Kepler appears not to have been convinced by this analogy, as we may deduce from his comment in the *Conversation*. Being from Bohemia himself, he was well acquainted with its topological characteristics and must have realized that the circular shape of the mountain range as it was represented on maps was exaggerated and that Bohemia could therefore not hold as an analogous feature to the circular phenomenon on the Moon.[52] Moreover, when Kepler read the *Sidereus*

Geometry, 243-44, my addition) While I agree with the point that Galileo needed the crater to be proportionally bigger than it is in reality to allow for the comparison with the topography of Bohemia, I cannot find any indication in Galileo's argument for his wanting to convince his readers that the Moon is covered all over with features similar to Bohemia. The question disregarded by Edgerton is *why* Galileo wanted to draw a comparison to Bohemia as well as the implications of this question.

50 Scott Montgomery also argues that Galileo distorted features on the Moon so as "to make the Moon look more Earth-like than it was, removing the most alien features and contouring others in accord with certain conventions of geographic representations for maps." He additionally points out that "the prominent 'explosive' appearance of several craters (Tycho, Copernicus, Kepler) is missing entirely, but would have been plainly visible, given the number and resolution of other craters Galileo drew." Yet although Montgomery mentions the exaggerated size of the central crater and points out that Galileo distorted "the irregularity of the terminator to which he gave an excessively scalloped appearance," he does not note the fact that Galileo *differentiated* between Albategnius (the feature he compares to Bohemia) and the other features on the terminator, thus turning them into *different* features. Neither does he mention what I think is a crucial point: that Galileo had not made this differentiation in his own drawings, where several features on the terminator look crater-like. This leads Montgomery to conclude that Galileo's alterations were "done not necessarily at a conscious level, but as a consequence of his essential ideas about the Moon." (*Scientific Voice*, 225-26)

51 Admittedly, these modern editions of Ptolemy's *Geographia* only remotely relied on the original work, which had only survived in spurious twelfth- and thirteenth-century manuscripts anyway. The passage on Bohemia as well as its cartographic depiction are a modern addition. Nevertheless, the modern author-editors of the *Geographia* presented their works as translations of Ptolemy's classic and the new additions as a mere enhancement. They thus use Ptolemy's authority just like Galileo.

52 In one of the notes on the "Geographical Appendix" in the *Somnium*, he offers an explanation for the formation of Bohemia: He describes how over a long period of time the river Elbe washed away the softer soil, leaving behind the rocks, once covered with earth, that now are the mountain range of Bohemia. (*Kepler's Somnium*, 156)

nuncius, he verified Galileo's claims with a telescope which, together with a copy of the book, had been presented to his employer, the Emperor Rudolf II. On observing the lunar surface through the telescope, Kepler too saw several of these large circular phenomena rather than only one. This made it unlikely that the feature pointed out by Galileo was equivalent to Bohemia. Yet Kepler was also intrigued by its perfect circularity and vast size. He offered his own interpretation, suggesting that it may be a huge circular fortress built by the inhabitants of the Moon to escape the calamities of the lunar climate.[53] We will return to this interpretation in the next chapter; what is interesting in this context is that Kepler devised an explanation which would not only account for the remarkable circularity of the phenomenon, but allow for more than one of its kind. Accordingly, in the *Somnium* Kepler refers to the "towns on the Moon."

Besides images representing what Galileo claims to be analogous phenomena on Earth and which serve both an explanatory and an illustrative purpose, Galileo at times uses images exclusively for their pictorial value. While such images do not confer meaning onto the visual image produced by the telescope, they may help to collect seemingly random spots into meaningful images (images, that is, with a meaning of their own). Such images make the telescopic image memorable as well as recognizable as a pattern. Accordingly, they assure that when Galileo's observations are repeated by someone else, the observer will see the spots in the kind of pattern Galileo desired and with as much precision as possible. In view of the poor quality of the early telescopes, this was an important issue. In addition, such images enhance the beauty of the scene, or rather they beautify a visual impression which to most of Galileo's contemporaries must have looked shockingly ugly.[54] In referring to the many dark spots that are seen in the lower horn of the new Moon Galileo remarks:

> This lunar surface, which is decorated with spots like dark blue eyes in the tail of a peacock, is rendered similar to those small glass vessels which, plunged into cold water while still warm, crack and acquire a wavy surface, after which they are commonly called ice-glasses.
>
> Haec lunaris superficies, qua maculis, instar pavonis caudae caeruleis oculis, distinguitur, vitreis illis vasculis redditur consimilis, quae adhuc calentia in frigidam immissa, perfractam undosamque superficiem acquirunt, ex quo a vulgo glaciales Cyathi nuncupantur.[55]

---

[53] Kepler, *Conversation*, 27-28. Apparently not quite comfortable with making such a bold proposition, Kepler argues that he is taking up a point made by Plutarch. See also the "Geographical Appendix" to the *Dream*, 151, where he gives a detailed description of how these "towns on the Moon" are built. In the *Conversation*, Kepler moreover criticizes Galileo for only depicting round bowl-like valleys on the Moon. Kepler holds that there cannot only be deep, round valleys without there also being long, steep ones like we find on Earth, which are formed by rivers running through them: "Why then do you report no such long spots on the moon?" In Kepler's view, Galileo's description is unconvincing for its lack of similarity to the Earth. He then advances a tentative guess suggesting that the Moon could be "like pumic, with very many pores of the largest size opening up on all sides" – and thereby finds another way to preserve the kinship between Earth and Moon. (25)

[54] See Kepler's comment quoted in note 24, *supra*.

[55] *SN*, 43, and *Opere*, vol. 3, pt. 1, 65.

The image of the peacock's tail, which Galileo evokes first, clearly serves to embellish the hardly very attractive idea of a lunar face full of cracks, excrescences and spots. It thus reinstates the Moon in its celestial beauty, this time together with the Earth. Galileo's purpose in emphasizing 'the relationship and similarity between the Moon and Earth' was to raise the Earth into the noble ranks of the heavens and not to tumble the whole universe into terrestrial chaos.[56] Thus his announcement of a future project (finally to become the *Dialogue*): "For we will demonstrate that she [i.e., the Earth] is movable and surpasses the Moon in brightness, and that she is not the dump heap of filth and dregs of the universe, and we will confirm this with innumerable arguments from nature."[57]

Besides gracing the spotted Moon with finery, the image of the eyes on a peacock's tail serves as a memorable image of illustration. It is hard to think of a more felicitous image than that of the peculiar design of the 'dark blue eyes' in a peacock's tail to evoke the pattern of light and shadow that gives prominence to the bowl-like indentures as they are obliquely hit by the sunlight at new Moon: "the dark part of the spot always faces the direction of the Sun and the brighter border surrounds the dark spot on the side turned away from the Sun and facing the dark part of the Moon."[58] Galileo here offers his readers an image that they can easily remember and which at the same time accurately reproduces the luminous rims surrounding the dark spots on the telescopic image.

With the second comparison, Galileo strikes a homelier note especially to his Venetian audience, familiar with the various glass-manufacturing techniques the *Serenissima* was so famous for. Eileen Reeves explains that 'ice-glass' (*glaciales Cyathi*) "represented a kind of Mannerist development in glass-making: Bored with the limpid purity of crystal, the Venetian artisans ... began producing a material that was rough, covered with innumerable fractures, and opaque." It was the latest fashion among the very rich who prided themselves on their sophisticated tastes. Reeves observes that images like this were meant to be of aesthetic appeal to readers. They were to suggest that "the unending chain of peaks, valleys, and craters, like that on fine Venetian glassware, are not a random production of nature but a pattern imposed by that *Miglior Fabbro*, the Best of Artisans."[59]

Galileo's analogy to Venetian glassware is no doubt ingenious for the way it conveys an aura of elegance and adds the notion of surface texture to the exclusively pictorial quality of the peacock image. Yet if we consider the fact that, in order to convince his contemporaries of the validity of his argument, Galileo primarily needed to convince them of the telescope's reliability as an instrument, the use of an illustra-

---

[56] On this crucial chapter in the history of ideas, along with Nicolson's pioneering work, see Hans Blumenberg, *Paradigmen*. Blumenberg speaks of a "stellarization of the Earth" (115). For an argument concerning Galileo's rejection of the hierarchization of space, see Michel Foucault, "Of Other Spaces," English translation in *Diacritics* 16, no. 1 (1986): 22-27.

[57] *SN*, 57. Explanation mine. – "vagam enim illam ac Lunam splendore superantem, non autem sordium mundanarumque fecum sentinam, esse demonstrabimus, et naturalibus quoque rationibus sexcentis confirmabimus." (*Opere*, vol. 3, pt. 1, 75)

[58] *Ibid.*, 43. For the original text, see p. 48, *supra*.

[59] *Painting the Heaven*, 150-51.

tive image involving impure glass seems a rather odd choice. Instead of being read as an image for the spots on the surface of the Moon, the impurities in the ice-glass could be brought into an unintended analogy with impurities in the lenses of Galileo's telescope. Whereas in the former case such impurities were desired, in the production of lenses they were an unwelcome, but often inevitable effect of the cooling process of the glass. The image of ice-glass thus entails the risk of drawing attention to the artificiality of the telescopic vision, to the fact that the telescope was a man-made instrument, which may not be trustworthy.[60]

The traditional Aristotelian distinction between the natural and the artificial and the notion that only natural sense perception is acceptable for the establishment of evidence made the acceptability of evidence produced with an artificially fabricated instrument like the telescope problematic.[61] And the firm bond between the visual and the cognitive, so central to Galileo's argument in the *Sidereus nuncius*, can only be confirmed if it is guaranteed that the telescope furnishes trustworthy images. Yet the way Galileo chooses to establish this bond is by rhetorical artifice, by establishing artificial bonds between the world out there and the world down here, and these tell a different story. Galileo's use of imagery is ingenious for how it serves to translate the unfamiliar appearance of the Moon in familiar terms, yet it also has its limitations in that its language is too much implicated in this world and carries with it accidental meanings, at times of an unwelcome sort.

A discussion of Galileo's use of imagery in the *Sidereus nuncius* is incomplete without an examination of the striking emblem devised in the dedication. At the beginning of the book, Galileo placed a dedicatory letter to his future patron, Cosimo II de' Medici, the new Grand Duke of Tuscany. In this letter he presents the satellites of Jupiter – or Medicean stars,[62] as he decided to call them – as a divine gift to Cosimo and as an emblem devised by God to celebrate, sanction and confirm the immortality of the Medici rule.[63] This whole dedication and its rhetorical presentation is of great

60 As Mario Biagioli points out, "the early challenges and Galileo's related responses focused on the very existence of the things he claimed to have observed rather than on their Copernican significance. The reliability of the telescope, rather than the philosophical or theological plausibility of the Copernican system, was the main target of Galileo's adversaries and the subject of his replies in the period immediately following the publication of the *Sidereus nuncius*." (*Galileo Courtier*, 95) For a wider discussion of the history, construction and reliability of early telescopes, see Albert Van Helden, "The Invention of the Telescope," in *Transactions of the American Philosophical Society* 67, no. 4 (1977): 5-67.

61 For a discussion of the redefinition of the notion of experience in the early seventeenth century, see Peter Dear, "Jesuit Mathematical Science and the Reconstitution of Experience in the Early Seventeenth Century," in *Studies in History and Philosophy of Science* 18 (1987): 133-75, and *idem*, *Discipline & Experience: The Mathematical Way in the Scientific Revolution* (Chicago: University of Chicago Press, 1995), as well as *idem*, *Revolutionizing the Sciences: European Knowledge and Its Ambitions, 1500-1700* (London: Pelgrave, 2001), 66-79. See also Simon Schaffer's discussion of the concept of evidence and its relation, on the one hand, to the person capable of giving evidence and, on the other hand, to the machine. Cf. "Self-Evidence," in *Critical Inquiry* 19 (1992): 327-62.

62 It was customary to also call planets and satellites by the name of 'stars'.

63 For a detailed discussion of the ingenious way in this emblem took up and reworked an association between Cosimo I and Jupiter that was already established, see Mario Biagioli, "Galileo the

interest in our context. The dedication is introduced with a passage describing the efforts of those who attempted to secure the great deeds of eminent figures in the memory of all times by devising emblems for their greatness. Thwarted by the inevitable corruption of all earthly things, these commendable men eventually turned to "the eternal orbs of the most brilliant stars" in search of "more incorruptible symbols" for earthly fame. They named the stars after "those who, because of their illustrious and almost divine exploits, were judged worthy to enjoy with the stars an eternal life."[64] Unfortunately, this ingenious human invention has long been out of use, since all the stars were already occupied with names. Even the attempt made by Augustus, who sought to glorify Julius Cesar by giving his name to a new star that had appeared at the time of Cesar's death, was thwarted by the comet's extinction shortly afterwards. But now Galileo is renewing this tradition by naming the satellites of Jupiter he recently discovered after the glorious Medici.[65]

When Galileo inscribes the name of the Medici into the stars, this act suspiciously parallels Ariosto's erecting a temple of earthly Fame on the Moon. Galileo gives the screw another turn by claiming that in this singular case it is by the virtue of the great Cosimo alone that these stars will be imparted the immortality of their name. But then what is in a name? We saw what Ariosto in his allegory on Fame had to say about the veracity of the praise authors of great works of literature lavish on their powerful patrons. Galileo used the lunar allegory by the poet he most cherished as a source of inspiration for his dedication,[66] not to disparage the man he wanted to become his patron, but to take some critical distance from the conventions of emblematics and its anthropocentric imagery, so at odds with the astronomical message his book contains. This has important consequences for the imagery he uses in presenting this message and its heavily anthropocentric nature. We should therefore read Galileo's use of imagery in the light of his implicit criticism of celebratory emblematics. We should then consider both his use of imagery and his ironic distancing from it in the context of his general ideas regarding the possibilities and limitations of natural philosophy. This will help us to find out more about the role of imagery in Galileo's writings and the extent of its legitimacy.

What Mario Biagioli calls the "translati[on] of scientific *mirabilia* into the discourse of the court",[67] and what I am here referring to more generally as the rhetoric of imagery of Galileo's dedication, represents a glaring discrepancy to the scientific

---

Emblem Maker," and *idem*, *Galileo Courtier*, chapters I and II. Jean Dietz Moss analyzes Galileo's dedicatory letter to Cosimo with regard to the way it employs elements of classical rhetoric. (*Novelties*, 78-80)

64 *SN*, 30. – "attamen longius illa prospiciens, cum optime intelligeret, omnia humana monumenta vi tempestate ac vetustate tandem interire, incorruptiora signa excogitavit, in quae tempus edax atque invidiosa vetustas nullum sibi ius vidicaret. In caelum itaque migrans, clarissimorum Siderum notis sempiternis illis Orbibus eorum nomina consignavit, qui ob egregia ac prope divina facinora digni habiti sunt, qui una cum Astris aevo sempiterno fruerentur." (*Opere*, vol. 3, pt. 1, 55)

65 *SN*, 29-30.

66 For Galileo's admiration for Ariosto and his notorious preference for Ariosto over Tasso, see Galileo "Postille all'Ariosto," in *Opere*, vol. 9, 149-94, and "Considerazioni al Tasso," *ibid.*, 59-148.

67 "Galileo the Emblem Maker," 242.

issues of the *Sidereus nuncius*. In the first place, the notion of rescuing human fame from the corruptibility of earthly things by securing it in the permanence of the stars makes an odd appearance in the context of Galileo's commitment to tumbling the Aristotelian dichotomy between the unchanging and perfect heavens and an Earth that is subject to continual degeneration and corruption. Also, the very geocentric vision of Jupiter watching over his Medicean protégé, sending his rays to make Cosimo a powerful ruler, is hardly compatible with Galileo's cosmological argument for Copernicanism, which deprives the Earth of its central position and significance in the cosmos. On the other hand, the interpolation recounting Augustus' failed attempt to dedicate a new star to Caesar seems to contradict the concept of heavenly permanence and stands out awkwardly against Galileo's own dedication of 'Medicean stars'. The entire rhetoric of Galileo's emblematic narrative thus stands in a deeply ironic relation to the scientific propositions made in the *Sidereus nuncius*. Now what should we make of this?

Obviously, the rules of dedicatory emblematics were hardly compatible with 'straight' scientific issues. Yet this is precisely the point: Galileo's scientific claims lay bare the emblem-making as a mere set of conventions, a game that has nothing to do with the content of Galileo's message, but is only form. At the same time, it becomes clear how little 'straight' Galileo's sidereal message itself is and how much it depends on the meandering course of poetry when it comes to binding the astronomical issues into a narrative that has the power to persuade. The courtly rhetoric in the dedication becomes a prototypical example for the kind of appropriative, self-centered discourse humans create in order to make their surroundings meaningful in ways that cater to their interests. Similarly, images that are used for explaining the contents of the actual message and for rendering the strange world it describes visible in the readers' imagination reveal an effort to make what is strange meaningful within the limited frame of human perception. Just as Ariosto's Moon becomes a faithful mirror which faithfully renders all the terrestrial lies, so the use of imagery in the *Sidereus nuncius* becomes a mirror of the rules and conventions of the self-referential system of the Medici court and, more broadly, of human conventions of representation in general.

Are we to conclude from this that the techniques of representation in the *Sidereus nuncius* are entirely illegitimate? The short answer to this question is: not entirely, as long as we take Galileo's images at face value only. The longer one will be given in the rest of this chapter.

In introducing his dedicatory emblem of the Medicean stars, Galileo makes two references to the peculiarly human preoccupation with and need for images. The first such reference appears right in the first sentence, but since it depends on how the word '*humanitas*' is translated, it is generally overlooked: when referring to the endeavor of those who seek to preserve the names of excellent men from oblivion, Galileo calls it "Praeclarum sane atque humanitatis plenum ... institutum"; this phrase is usually translated with "A most excellent and *kind* service" (my emphasis), interpreting the phrase 'humanitatis plenum' as 'kind' or 'humane' in accordance with its immediate context. However, the phrase can, of course, also mean 'full of humanity' or simply 'typically human'; and in view of the gross inconsistencies between the dedicatory narrative and the astronomical message that follows it, this secondary meaning should not just be dismissed. Accordingly, Galileo calls his own endeavor to project the name of the Medici into the sky *typically human*.

After enumerating several ways in which eminent men have been commemorated, Galileo explains: "For such is the condition of the human mind that unless continuously struck by images of things ('*simulacra*') rushing into it from the outside, all memories easily escape from it."[68] In the first place, Galileo is here emphasizing the importance of finding suitable repositories to ensure the long-lasting commemoration of the great and counteract human forgetfulness. Yet the phrasing of his argument is strangely specific: rather than just referring to a typically human trait, he points to the particular 'condition of the human mind'. Rather than just as a reference to the emblematics of the dedication, we should read this observation in the context of Galileo's preoccupation with using felicitous as well as striking imagery in presenting his astronomical argument. Read in this context, Galileo's observation seems to refer to a more specific characteristic of the human mind: its need to visualize the information it obtains. The visualizing of information not only helps the mind to absorb and remember this information, but also to understand it.

We have seen earlier how Galileo's use of imagery caters to the reader's visual imagination by supplying it with memorable as well as meaningful images. The word used for 'images of things' in the original version of this passage is '*simulacra*'. As Isabelle Pantin explains, the theory of *simulacra* was developed by Lucretius in *De natura rerum*, IV, where *simulacra* are described as membranes or subtle shells that are emitted by objects and spring from their surface in order to roam about in space maintaining their forms. In Lucretius, this phenomenon serves as an explanation for all phenomena pertaining to sensation, the imagination and memory. Pantin argues that Galileo does not give the concept of *simulacra* the precise meaning it has in Lucretius, and with regard to the more obvious reference of his observation she is certainly right. Yet considering the kind of images Galileo uses in describing the lunar surface and their specific function, these images quite accurately correspond to the Lucretian concept of *simulacra*: empty shells or images without substance that serve to feed the readers' imagination so as to enable them to visualize what Galileo describes by projecting images of familiar objects onto a strange vision.[69]

If the human mind, in order to be able to imagine something strange, needs to be fed with empty images of familiar provenance, and if Galileo uses such images in describing the phenomena he perceived through his telescope, does this imply that the application of his images is invalid and their descriptive power mere rhetorical trickery? If such were categorically the case, we would have to agree with the Aristotelians in saying that anything beyond the range of our terrestrial experience is beyond comparison and for the most part beyond knowledge. How then is the legitimacy of Galileo's images defined? As I will try to show, it lies in accepting that these images refer to appearances only rather than to the essences behind the appearances. The passage relevant in this connection includes both a mocking criticism of Aristotelian

---

68 *SN*, 29. – "Eiusmodi est enim humanae mentis conditio, ut nisi assiduis rerum simulacris in eam extrinsecus irrumpentibus pulsetur, omnis ex illa recordatio facile effluat." (*Opere*, vol. 3, pt. 1, 55)

69 *Le méssager céleste*, 51, note 12.

narrow-mindedness and a striking reference to painting. Galileo derides the Aristotelians for attempting to explain the universe exclusively on the basis of Aristotle's writings:

> [Some philosophers] wish never to raise their eyes from those pages – as if this great book of the universe had been written to be read by nobody but Aristotle. ... These fellows ... put me in mind of certain capricious painters who occasionally constrain themselves, for sport, to represent a human face or something else by throwing together now some agricultural implements, again some fruits, or perhaps the flowers of this or that season. Such bizarre actions, so long as they are proposed in jest, are both pretty and pleasant, and reveal greater resourcefulness in some artists than in others. ... But if anyone ... should then conclude in general that every other manner of representation was blameworthy and imperfect, it is certain that Cigoli and other illustrious painters would laugh him to scorn.[70]

> alcuni severi difensori di ogni minuzia peripatetica ... non vogliono mai sollevar gli occhi da quelle carte, quasi che questo gran libro del mondo non fosse scritto dalla natura per esser letto da altri che da Aristotele, ... Questi ... mi fanno sovvenire di certi obblighi a i quali tal volta per ischerzo si astringono capricciosi pittori, di voler rappresentare un volto umano o altra figura con l'accozzamento ora de'soli strumenti dell'agricoltura, ora de' frutti solamente o de i fiori di questa o di quella stagione: le quali bizzarrie, sin che vengono proposte per ischerzo, son belle e piacevoli, e mostrano maggior perspicacità in questo artefice che in quello, ... ma se alcuno, per aver forse consumati tutti i suoi studii in simil foggia di dipingere, volesse poi universalmente concludere, ogni altra maniere d'imitare esser imperfetta e biasimevole, certo che 'l Cigoli e gli altri pittori illustri si riderebbono di lui.[71]

At the beginning of this extract the Aristotelians are criticized for confusing books: Instead of studying the 'great book of the universe', they study the writings of Aristotle and, by projecting Aristotle's theories onto the universe, pretend that this *is* the book of the universe. We will return to this point later on in connection with another reference to reading the book of the universe. For the moment I would like to turn to the kind of paintings to which the Aristotelian projections are compared. In these 'bizarre performances' (*bizzarrie*) an object is represented through images of other objects that bear no intrinsic relation to what they represent; instead, the indirect representation is achieved through the combination and setting in perspective of outward images only. It is remarkable how, more than anything else, this technique evokes Galileo's own use of imagery in the *Sidereus nuncius*. This is most obviously the case in the extravagant imagery of the dedication. Yet it equally applies with respect to the images that illustrate Galileo's astronomical argument, for is Galileo there not also using the images of familiar objects or phenomena to describe something else? And even while one might argue that some of Galileo's analogical comparisons assert more essential similarities between terrestrial and lunar phenomena than those we find between a potato and a human nose, are not these too still projections of one reality onto another?

---

[70] Quoted in Stillman Drake, ed., *Discoveries and Opinions of Galileo* (Garden City, NJ: Doubleday, 1957), 127. Addition mine. The passage is taken from the *Third Letter on Sunspots*.

[71] *Opere*, vol. 5, 190-91.

The crucial redeeming distinction in the case of the painters is that their performances are 'proposed in jest' (*proposte per ischerzo*), and in a way this is also true in Galileo's case, even though he might have dedicated himself to this 'sport' less freely than the painters. According to Galileo, the essential distinction between the Aristotelians and himself lies in the fact that, unlike them, he is not "satisfied with worshiping shadows" (*content*[*o*] ... *di adorar l'ombre*) that he takes for true substances, but recognizes them for what they are: merely an adumbration of a thing of substance.[72] And it is here that Galileo's images acquire their legitimacy. Galileo's astronomical argument refers to appearances and not to the essential nature of terrestrial or lunar substances. When he is pointing out similarities between the Moon and the Earth, he is not defining the essence of lunar substance, nor is he saying that the essence of lunar substance equals that of terrestrial substance. All he is saying is that they resemble each other because of the correspondence of some of their distinctive characteristics and that these characteristics can be deduced from certain phenomena on the Moon and their close resemblance to phenomena on Earth. Galileo's argument is not about the essence of things, but about the way they appear to the human observer. Indeed, he was convinced that it was impossible for a human being to know about the true essences of things. What the student of nature could do was attempt to define some distinctive characteristics of things, and by comparing them deduce affinities between substances. He presents his opinion clearly in the following extract:

> all the things among which men wander remain equally unknown, and we pass by things both near and remote with very little or no real acquisition of knowledge. When I ask what the substance of clouds may be, and am told it is a moist vapor, I shall wish to know in turn what vapor is. Peradventure I shall be told that it is water, which when attenuated by heat is resolved into vapor. Equally curious about what water is, I shall then seek to find this out, ultimately learning that it is this fluid body which runs in our rivers and which we constantly handle. But this final information about water is no more intimate than what I knew about clouds in the first place; it is merely closer at hand and dependent on more of the senses. In the same way I know no more about the true essences of earth and fire than about those of the moon or the sun ...
>
> But if what we wish to fix in our mind is the apprehending of some properties of things [*affezioni*], then it seems to me that we need not despair of our ability to acquire this respecting distant bodies as well as those close at hand – and perhaps in some cases even more precisely in the former than in the latter. Who does not understand the periods and movements of the planets better than those of the waters in our various oceans? Was not the spherical shape of the moon discovered long before that of the earth, and much more easily? Is it not still argued whether the earth rest motionless or goes wandering, whereas we definitely know the movements of many stars?[73]

[72] *Opere*, vol. 7, 30. Adaptation mine. For the reference to the Aristotelians as shadow worshipers, see Galileo's preface to the *Dialogue*.

[73] Quoted from Stillman Drake, *Discoveries*, 123-24. Emphases and addition mine. The passage appears in Galileo's third letter on sunspots of 1612. For Galileo's cautious program and limited objectives for science, cf. Stillman Drake, "Galileo's Language: Mathematics and Poetry in a New Science," in *Yale French Studies* 49 (1973): 13-27.

nè veggo che nell'intender queste sostanze vicine aviamo altro vantaggio che la copia de' particolari, ma tutti egualmente ignoti, per i quali andiamo vagando, trapassando con pochissimo o niuno acquisto dall'uno all'altro. E se, domandando io qual sia la sustanza delle nugole, mi sarà detto che è un vapore umido, io di nuovo desidererò sapere che cosa sia il vapore; mi sarà per avventura insegnato, esser acqua, per virtù del caldo attenuata, ed in quello resoluta; ma io, egualmente dubbioso di ciò che sia l'acqua, ricercandolo, intenderò finalmente, esser quel corpo fluido che scorre per i fiumi e che noi continuamente maneggiamo e trattiamo: ma tal notizia dell'acqua è solamente più vicina e dependente da più sensi, ma non più intrinseca di quella che io avevo per avanti delle nugole. E nell'istesso modo non più intendo della vera essenza della terra o del fuoco, che della Luna o del Sole ...

Ma se vorremo fermarci nell'apprensione di alcune affezioni, non mi par che sia da desperar di poter conseguirle anco ne i corpi lontanissimi da noi, non meno che ne i prossimi, anzi tal una per aventura più esattamente in quelli che in questi. E chi non intende meglio i periodi de i movimenti de i pianeti, che quelli dell'acque di diversi mari? chi non sa che molto prima e più speditamente fu compresa la figura sferica nel corpo lunare che nel terrestre? e non è egli ancora controverso se l'istessa Terra resti immobile o pur vadia vagando, mentre che noi siamo certissimi de i movimenti di non poche stelle?[74]

The translation of Galileo's '*affezioni*' as 'properties' is not quite accurate since it leaves out the connotation of subjectivity present in the original: *affezioni*, rather than denoting properties of things themselves, that is, *essential* properties, refers to attributes we give to things based on our subjective impression of them. Such attributes or characteristics (again the metaphorical image of the face comes to mind) are determined by the way things 'affect' us, by their appearance to the human observer.[75]

Yet if Galileo does not define lunar substance and merely refers to appearances, what then is his point against the Aristotelians and their principle of a radical dichotomy between celestial and terrestrial substances? Galileo's main point against this proposition is, first of all, that the Aristotelians had no evidence for claiming the existence of a substantial difference between celestial and terrestrial matter other than appearances; and, secondly, that if appearances could be shown to be other than what the Aristotelians claimed, i.e., similar rather than different, there was no justification for such a claim and no obstacle to setting up the counterclaim of a substantial affinity

---

[74] *Opere*, vol. 5, 187-88.

[75] *Opere,* vol. 5, 188. According to one dictionary, the word *affezione* can also signify a subjective sensation. This meaning is illustrated with the following quotation from Galileo: "quell'affezione grata o molesta che noi abbiamo nominato caldo, bruciore o scottamento," i.e., 'this *affezione*, pleasant or cumbersome, that we have called hot, scalding or burning'. Cf. *Vocabolario della lingua italiana* (Rome: Istituto dell'enciclopedia italiana, 1986). Indeed, as Timothy Reiss has pointed out to me in correspondence, the Latin *affectiones/adfectiones* was commonly used as a synonym for *pathos* or *passiones*. This links Galileo's argument to the vast and complex theory of passions so important in Galileo's day, thus strengthening the argument of subjectivity. In his history of the theory of passions, Robert C. Solomon explains that the ancients often treated passion as a kind of madness, and the Stoics diagnosed the passions as profound and often fatal misunderstandings; early modern philosophers have treated them as confused perceptions and distorted judgments. Cf. *The Passions: Emotions and the Meaning of Life*. (Indianapolis: Hackett, 1993).

between terrestrial and lunar substances – even if such a claim may in turn have to be qualified on the ground of new, diverging appearances.

The other basic argument against the Aristotelians is that they, like everyone else, were essentially ignorant about terrestrial substances.[76] As Galileo makes clear, knowledge of the terrestrial world, while often more detailed than that of remote objects, was deficient in instances where it was possible to characterize the latter more clearly. In pointing out such instances, Galileo alienates his readers from the familiar world by emphasizing that acquiring knowledge about terrestrial objects is in many ways as complex and endless a process as acquiring knowledge about distant objects. Hence, our physical proximity to them does not guarantee that we are familiar with even the most basic of their characteristics. The Aristotelian dichotomy is therefore untenable even with respect to the obtainability of knowledge.[77]

The principle that it is impossible for human beings to penetrate to the essence of things has direct consequences for the descriptive images Galileo uses to compare the telescopic visions to. For if our knowledge about things is limited to appearances, it follows that the relation between a metaphoric image and the appearance it describes differs in degree only from the relation between our mental conception of an object and this object as it is in its essence. In neither relation is the one identical with the other. The degree of perceptible difference between the two elements of the first relation lies in the number of characteristics in which they coincide and differ from each other. Accordingly, Galileo's use of descriptive images is a legitimate means of illustration as long as these images do not matter, that is, as long as they are not taken for the thing they customarily represent but are used as *simulacra*, i.e., as shells without substance that share characteristics with the *simulacra* they describe. Used in this way, such images serve as expedients that permit Galileo to make the strange imaginable to his readers.

However, much of the force and charm of Galileo's rhetoric of imagery lies, of course, in transgressing the bounds of legitimacy. In some cases Galileo exploits additional connotations of an image which do not participate in the characteristics shared between the image and what it refers to (e.g., the image of the peacock's tail, which besides providing an accurate pictorial image lends a sense of beauty to the telescopic image). In other places, he introduces images which bear no apparent relation to what they refer to but turn the strange world seen through the telescope into a metaphorical representation of a domestic image rather than the other way round (e.g., the image of kinship between Moon and Earth). In such instances Galileo's images partake of the games and fictions of poetry and are, as it were, proposed in jest. We have seen how such games can be useful to ends like emotively affecting and persuading the public addressed or making the strange accessible to the imagination, but they deviate from the serious study of nature by misrepresenting the information it offers.

---

[76] Maurice Slawinski observes that Galileo here uses a "dialectical method which, effectively, founds a syllogism upon litotes (a rhetorical figure which consists in asserting something by denying the opposite)." ("Rhetoric in Science," 93)

[77] For Galileo's distinction between "name and thing, word and essence," see also Brian Vickers, "Epideictic Rhetoric," 88.

Yet no matter whether images are used licitly or illicitly, the language of images can never serve as a neutral tool to study and discourse about nature since it carries the dross of illegitimate identities and always entails the risk of deviating into sense. This is why Galileo looked elsewhere for a language which, while immediately referring to the sensible world, only involved isolated characteristics abstracted from sensible objects and thus ignored the identities of these objects where they were of no concern. These requirements were met by mathematics, which Galileo perceived as the key to the acquisition of knowledge about the universe. Let me quote once again the famous passage in which this is declared, this time with the illuminating juxtaposition Galileo uses to introduce it:

> Possibly he [i.e., Signor Sarsi] thinks that philosophy is a book of fiction created by some man, like the *Iliad* or *Orlando Furioso* – books in which the least important thing is whether what is written in them is true. Well, Sig. Sarsi, that is not the way matters stand. Philosophy is written in this grand book – I mean the universe – which stands continually open to our gaze, but it cannot be understood unless one first learns to comprehend the language and interpret the characters in which it is written. It is written in the language of mathematics, and its characters are triangles, circles, and other geometrical figures, without which it is humanly impossible to understand a single word of it; without these, one is wandering about in a dark labyrinth.[78]

> e forse stima [i.e., il Signor Sarsi] che la filosofia sia un libro e una fantasia d'un uomo, come l'Iliade e l'Orlando Furioso, libri ne' quali la meno importante cosa è che quello che vi è scritto sia vero. Sig. Sarsi, la cosa non istà così. La filosofia è scritta in questo grandissimo libro che continuamente ci sta aperto innanzi a gli occhi (io dico l'universo), ma non si può intendere se prima non s'impara a intender la lingua, e conoscer i caratteri, ne' quali è scritto. Egli è scritto in lingua matematica, e i caratteri son triangoli, cerchi, ed altre figure geometriche, senza i quali mezi è impossibile a intenderne umanamente parola; senza questi è un aggirarsi vanamente per un oscuro laberinto."[79]

Stillman Drake has focused on the basic distinction of three elements in this metaphor. "The three elements are: a certain book, what is written in it, and the language in which that is written there." He explains that the book is the sensible world, what is written in that book is proper philosophy (proper because it deals with the sensible world), while the language in which proper philosophy is written is the language of mathematics. "This language is identified by Galileo neither with the sensible world nor with philosophy; still less is it treated as anything worthy of study for its own sake. Rather, Galileo speaks of the language of mathematics as the unique means to an understanding of something else."[80] The philosophy written in the book of the universe is the knowledge about nature that we may acquire through studying the sensible world. This philosophy is not a fictitious fantasy, but immediately refers to the sensible world open to our gaze. Still, it is not tantamount to what the book contains, that is, the

---

[78] "The Assayer" in *The Controversy on the Comets of 1618*, ed. and trans. Stillman Drake and C. D. O'Mailley (Philadelphia: University of Pennsylvania Press, 1960), 183-84.

[79] *Opere,* vol. 6, 232.

[80] "Galileo's Language," 17.

sensible world itself, but corresponds to what with our limited insight we may grasp in reference to it ('*intenderne umanamente*'). Mathematics is the key to this knowledge. Drake points out that Galileo's metaphor "presented mathematics as an instrument that would enable philosophy to discourse with accuracy about the sensible world."[81] At the same time, Galileo presents mathematics as a language that is already written in the book of the universe, that is universal and not man-made, although it can be learned and used by human beings. It is the perfect instrument for the study of the universe because it expresses those of its characteristics that are humanly understandable without imposing a limiting human vision on it. As opposed to the language of images, it is not anthropocentric and does not specify more than what it immediately concerns.[82]

Yet as desirable as such an objective mode of communication with nature might be, when it comes to formulating one's findings into a thesis or, even more, communicating them to someone else, the language of mathematics proves inadequate to the task. This requires a degree of specificity that only a topical language can offer. The triangles, circles and other geometrical figures that form the characters of mathematics must therefore be translated into the words and images of a man-made language. This will allow the human mind to interpret the mathematical information by transforming it into concepts that are imaginable as well as capable of persuading others of their validity.

While such interpretive and communicative acts inevitably involve elements of fiction and rhetorical artifice, they also serve to make manifest the close tie of the mathematical laws and relations to a material world connected to people's actual experiences.[83] They lend Galileo's argument an appearance of the real that is tailored to human fantasy, thus contrasting it to mere bookish theories completely unrelated to reality, like those which the Aristotelians cling to. At the same time, the ironic distance that Galileo maintains to his use of images makes room for a potential alternative mode

---

81 *Ibid.*, 18.

82 As Timothy Reiss has argued, "[f]or Galileo discourse represents not the object itself but the *distance* between the object and the mind *per*ceiving and then *con*ceiving it." In contrast to later philosophers, Galileo's conception of a discursive order does not rest on the premise of a relation of identity between the syntactic order of semiotic systems (whether mathematical or linguistic) and both the logical ordering of 'reason' and the structural organization of the world. Rather, Galileo "emphasizes that knowledge is a sign-manipulating activity," i.e., an activity that manipulates the phenomena of nature (interpreted as signs) into a relationship of analogy with the discursive signs (subject to the organization of the mind) which are used to describe them. (*Discourse*, 33-34) This applies to both mathematical and linguistic language; and, more specifically, the language of imagery: none of these languages represent the objects they refer to in themselves; they discourse *about* them. The only advantage mathematics has over the language of imagery is that it offers a neutral system of relations and does not superimpose false identities. In this sense, it can be said to inhere in nature. See also "Espaces," 23, where in relation to the passage from Galileo just quoted Reiss observes: "Mathematics does not signify, it is. It produces its own truth in the order of its own form." Translation mine.

83 Stillman Drake argues that the use of poetic devices served a most useful purpose for Galileo: "not indeed to represent anything, but to call vividly to mind some actual experience." (*Ibid.*, 20) While I agree with the latter part of his argument, I think there are many instances in the *Sidereus nuncius* where poetic devices are vital to representation.

of representation and for the possibility of things being different. Speaking in terms of Galileo's painterly analogy, the words and images he uses to describe the strange phenomena are not claimed to be the only possible way of representation, but they speak the language of the human mind, in need of images.

Surely it is no coincidence that Galileo introduces his metaphor of the book of the universe by differentiating it from another type of book, one that has no foundation in truth, but has sprung from human fantasy. This differentiation is also a juxtaposition. Indeed, it is a juxtaposition similar in kind to the one between the philosophy that may be read from the book of the universe written in the language of mathematics and the interpretation of this philosophy translated into a man-made language in Galileo's own books. The second kind of book, as much as it may attempt to render the true philosophy as accurately as possible, cannot help but be '*un libro e una fantasia d'uomo*'. I believe that, whether made explicitly or merely elusively suggested in the form of parody and allusion, Galileo's frequent references to fiction and poetic fantasies as well as his critical distancing from the distorted visions they offer make an important comment on the limitations as well as the potential of human inquiries about the universe. For while such inquiries are inevitably bound to a fixed place and time, to personal experiences and a limited perception, it is in their complex interconnections and the way in which a common object in our daily lives may unwind a tale about some remote place in the universe that they manifest their firm ties to reality and distinguish themselves from mere chimeras of the mind.

Perhaps the most suggestive metaphor for this is given in the relationship between theme, form and setting in the *Dialogue*: the setting is Sagredo's Venetian palazzo, from which, physically, the three interlocutors hardly budge. Yet their speculations revolve around the Moon and the Earth, the Sun and the planets and the stars in a journey so tempestuous that Salviati repeatedly expects to lose his orientation:

> I see we are once more going to engulf ourselves in a boundless sea from which there is no getting out, ever. This is navigating without compass, stars, oars, or rudder, in which we must needs either pass from bank to bank or run aground, or sail forever lost.
>
> Io veggo che noi torniamo di nuovo a ingolfarci in un pelago infinito da non ne uscir mai, perchè questo è un navigar senza bussola, senza stelle, senza remi, senza timone, onde convien per necessità o passare di scoglio in scoglio o dare in secco o navigar sempre per perduti.[84]

Several times during the four days of discussions, Salviati loses the thread of his argument and needs a cue from the others to get back on track.[85] This is not surprising when we consider the numerous digressions in the discussion and how frequently the interlocutors take recourse to extended analogies. Yet if the complex ramifications of the argument is at times confusing, this, as Salviati assures his interlocutors, is just as it should be:

---

[84] *D*, 44, and *Opere*, vol. 7, 69.

[85] Examples for this can be found on p. 57 and at the beginning of the second day.

> The digressions made up to this point are not so foreign to the matter in hand as to be called entirely apart from it. Moreover, from such things there result trains of reasoning awakened in the minds of not one of us alone, but all three. Besides, we are arguing for our own amusement, and are not obligated to any such strictness as one would be who was methodically treating a subject for professional reasons, with the intention of publishing it. I do not want this epic [*poema*] of ours to adhere so closely to poetic unity as to leave no room for episodes, for the introduction of which the slightest relevance ought to suffice. It should be almost as if we had met to tell stories [*quasi che noi ci fussimo radunati a contar favole*][86], so that it is permitted for me to relate anything which hearing yours may call to my mind.

> Le digressioni fatte sin qui non son talmente aliene dalla materia che si tratta, che si possan chiamar totalmente separate da quella; oltrechè dependono i ragionamenti da quelle cose che si vanno destrando per la fantasia non a un solo, ma a tre, che anco, di più, discorriamo per nostro gusto, nè siamo obligati a quella strettezza che sarebbe uno che *ex professo* trattasse metodicamente una materia, con intenzione anco di publicarla. Non voglio che il nostro poema si astringa tanto a quella unità, che non ci lasci campo aperto per gli episodii, per l'introduzion de' quali dovrà bastarci ogni piccolo attaccamento, e quasi che noi ci fussimo radunati a contar favole, quella sia lecito dire a me, che mi farà sovvenire il sentir la vostra.[87]

The universe is so vast and complex in its secret workings that its study cannot but raise endless chains of questions and problems that need to be solved. At the same time, the fundamental principle of a basic set of physical laws applicable throughout the universe, makes it possible to approach some of these questions by drawing comparisons to our more immediate surroundings. Galileo thus establishes the drawing of connections and digressing into comparable phenomena as a fundamental working method of natural philosophy. The accumulation of related processes into a vast collection of episodic stories about the workings of the universe most nearly represents the complexity such a philosophy faces.

By having Salviati refer to their dialogues as a *poema* that defies the rigid concept of unity and abounds in loosely interconnected episodes and digressions, Galileo calls to mind an epic poem like the *Orlando furioso*, where the confusing multiplicity of singular episodes finally reveals a pattern in which they all merge into harmony. Yet whereas such a poem has a plot which, however intricately devised, has its clear limits in the poet's mind and ends with the telling, the collection of episodes Salviati is creating with his friends is based on the infinitely more complex plot of the philosophy written in the book of the universe, of which they can only try to unravel a tiny fraction.

Their poem is, therefore, endlessly expandable and becomes more complex the more people are involved in its creation. Galileo not only firmly believed that the ramifications this plot offered were too complex to be fathomed by the enquiring mind of a single human philosopher, he also considered it likely that they were inexhaustible.[88]

---

[86] Here Galileo clearly alludes to Boccaccio's *Decameron*. Cf. Introduction, p. 17.

[87] *D*, 162, and *Opere*, vol. 7, 188.

[88] In the *Dialogue*, Sagredo refers to the extreme presumptuousness of those who want to make human abilities the measure of what nature is capable of performing: "On the contrary, there is not

He therefore saw the study of the universe as a continuous process that defies unity of method as well as completion and publication. For this reason it is best rendered in the open form of a dialogue. Any attempt to write a definitive description of the system of the universe would only result in a ludicrous parody of the one book, whole and continuously changing, the book of the universe. How very pitiful, therefore, a Simplicio, who carries his system of the universe in his pocket![89]

In contrast to the dizzying dimensions of their subject matter and Salviati's repeated fear of shipwreck, the tone of the argument stays strikingly contained while the systematic line of reasoning conveys a sense of complete control. Indeed, the argument evolves so logically that, when properly cued by Salviati, it seems to develop by itself. This effect is achieved through the figure of Sagredo, who at various points in the dialogue cuts Salviati short in his expositions and continues them himself.[90] It becomes even more powerful when through gradual interrogation the Aristotelian Simplicio is induced to develop an argument that radically contradicts all of his convictions. Thus, the vast complexity and inexhaustibility of the truths about the universe is juxtaposed with the logical connections between its different parts and the unambiguous nature of the conclusions one may draw from such connections. As vast and inexhaustible as the object of study is, such clarity will always guarantee that Salviati and his interlocutors do not get lost in the labyrinth of wild speculations, but will find their way back to the immediate reality of the Venetian palazzo. At the same time, it enables them to embark on their journey in the first place, even if it is metaphorical.

---

a single effect in nature, even the least that exists, such that the most ingenious theorists can arrive at a complete understanding of it." This presumption can only spring from never having understood anything: "For anyone who had experienced just once the perfect understanding of one single thing, and had truly tasted how knowledge is accomplished, would recognize that of the infinity of other truths he understands nothing." (101)

89 Speaking of Aristotle's *De caelo*, he says: "I keep it always in my pocket." (136)

90 A good example of this clever structural technique can be found when, cutting short Salviati's enumeration of similarities between Earth and Moon, Sagredo gives the explanation for the Moon's secondary light himself. (*D*, 67)

3

The Stuff that Dreams are Made of: Kepler's *Somnium*

At the end of his "Considerations on the Opinion of Copernicus" of 1615, Galileo argues for the legitimacy of an analogy originally introduced by Copernicus himself. The analogy illustrates the absence of contradiction between the appearance of a stable Earth and moving Sun and the reality of the contrary, by comparing our impression that the Sun moves and the Earth stands still to the impression we get aboard a ship that the shoreline recedes from us when in effect we are leaving it.[1] In his response to Paolo Antonio Foscarini's attempt to reconcile Copernican theory with the Sacred Scriptures, Cardinal Bellarmine had criticized Foscarini's employment of the analogy and contested its validity. Bellarmine had offered the evasive argument that in the case of the Sun there is no need to correct our impression of its movement since it is not based on an error; in the second case, we can correct our error by "seeing clearly that the ship moves and not the shore".[2] Galileo replied that "[t]he error of the apparent movement of the shore and stability of the ship is known by us after having many times observed the motion of boats from the shore, and many other times observed the shore from a boat; and so, if we could now stay on earth and now go to the Sun or other star, perhaps we would acquire sensible and certain knowledge of which one of them moves."[3] While thus defending the validity of the comparison, Galileo also concedes that there is a "great disparity between the small boat ... and the immense shore". He therefore proposes to alter the terms of the comparison by considering

> two bodies both of which are substantial and equally inclined toward motion. Thus it would be more relevant to compare between themselves two boats, in which case it is absolutely certain that the one we were on would always appear to us as mo-

1 Nicolaus Copernicus, *On the Revolutions*, book 1, in *Complete Works*, trans. Edward Rosen, ed. Jerzy Dobrzycki (Warsaw: Polish Scientific Publishers, 1978), vol. 2, 16.

2 Cardinal Bellarmine to Foscarini (12 April 1615) in Maurice Finocchiaro, *The Galileo Affair* (Berkeley: University of California Press, 1989), 67-69. – "vedendo chiaramente che la nave si muove e non il litto". (*Opere*, vol. 12, 171-72) The full title of Foscarini's book, published in Naples, conveys a good idea of its form and content: *Lettera sopra l'opinione de' Pittagorici e del Copernico della mobilità della terra e stabilità del sole e del nuovo Pittagorico sistema del mondo, nella quale si accordano ed appaciono i luoghi della Sacra Scrittura e le proposizioni teologiche, che giammai potessero addursi contro tale opinione*. Foscarini was Provincial of the Carmelites for Calabria as well as professor of theology at the University of Messina.

3 "L'errore della apparente mobilità del lito e stabilità della nave è conosciuto da noi doppo l'essere molte volte stati sopra 'l lito a osservare il moto delle barche, e molte altre in barca a osservare il lito: e così se potessimo ora stare in Terra ed ora andar nel Sole o in altra stella, forse verremmo in ogni cognizione sensata e sicura, qual di lor si muova". (*Ibid.*)

tionless, as long as we could not consider any other relationship but that which holds between these two ships.

There is, therefore, a very great need to correct the error about observing whether the earth or else the sun moves, for it is clear that to someone on the moon or any other planet it would always appear that it was standing still and the other stars were moving.[4]

la grandissima disparità che è tra una piccola barca, divisa da ogni suo ambiente, ed una spiaggia immensa, conosciuta da noi immobile per mille e mille esperienze, immobile, dico, rispetto all'acqua ed alla barca; è molto differente dal fare paragone tra due corpi, ambidue per sè consistenti e disposti egualmente al moto ed alla quiete: tal che meglio quadrerebbe il far paragone di due navi tra di loro, delle quali assolutamente ci parrebbe sempre stabile quella dove fussimo noi, tutta volta che non potessimo far altra relazione che quella che cade tra esse 2 navi.

Ci è, dunque, bisogno grandissimo di correggerr l'errore circa l'apparenza se la Terra o pure il Sole si muova, sendo chiaro che uno che fosse nella Luna o in qualsivoglia altro pianeta, sempre gli parrebbe di star fermo e che l'altre stelle si movessero.[5]

This is a clever move: Galileo first changes the terms of the comparison so as to avoid offending the Cardinal by directly contradicting him; he then offers strong evidence for its validity by introducing a change in perspective – for that the Moon and the planets move is clearly visible from the Earth. If we accept the claim that – just as on a boat – to 'someone on the moon or any other planet it would always appear that it was standing still', it becomes much more difficult to dismiss the analogy when it is applied to the Earth, Galileo's second boat, as it were. Finally, Galileo's adaptation of the analogy strengthens its argument of relativism by doing away with the absolute values of movement (boat) and stability (shore) and exchanging them for relative ones of movement *and* stability (two boats in relation to each other, both of which may either stand for a stable or a moving object). Yet with all the clever twists he gave to the argument one problem remained: in the absence of other empirical evidence, an observer would have to change position to one of the other 'boats' in order to determine whether the Earth does indeed move – a condition that was, obviously, impossible to meet.[6]

To change the angle of perspective to an extraterrestrial position, more specifically, to present the viewpoint of one of the other 'boats' in the heavens, that of the Moon, is just what Kepler attempts to do in his *Somnium*, or *Dream*, a work that occupied him from his student days in Tübingen until the day he died and which remained unfinished.[7] It is based on the principle that a living being on the Moon would be likely to develop an astronomy that takes the Moon as the stable center of the universe around which everything else moves, while the Earth serves as its moon. The Moon is thus a

4 "Galileo on the Copernican Opinion," in Finocchiaro, *The Galileo Affair*, 86.

5 *Opere*, vol. 5, 369-70.

6 Full proof of terrestrial motion was not available until the nineteenth century when Bessel discovered the stellar parallax and Foucault could demonstrate the earth's rotation by his experiments with the pendulum.

7 Timothy Reiss notes that if one were to measure the respective importance of Kepler's astronomical works in time then the lunar speculations of the *Somnium* would appear to occupy the greatest place in his researches. (*Discourse*, 143-44)

mirror that presents us with a mirror image of (faulty) terrestrial science.[8] The correct science is the result of the synthesis of the two faulty sciences. Besides offering a new perspective on the heavens, a second concern of the book is to imagine the nature of the lunar world itself, its climatic conditions and the forms of life such conditions might produce.

Both the cosmological and the selenographical argument of the *Dream* thus rely on the creation of experiences not available to Earth-dwellers. While the representation of these experiences is of fundamental importance to the theses Kepler presents in the *Dream*, the representation itself is obviously a fabrication, the experience of a lunar astronomy and topography a mere fiction of the imagination. The *Dream* uses this fiction as a rhetorical device. It does so in two ways: first of all, it uses it as truth by offering the kind of evidence necessary but not available in the form of a fictional narrative which presents the fiction as empirical evidence; and secondly, it uses it as a fiction, yet a fiction achieved primarily by translating Kepler's theory into a visual image with the help of the laws of optics and geometry.

The purpose of such a fiction is to make Kepler's theory accessible to the reader's imagination, to spell out the theory in visual terms, thus lending it credibility. Yet with all its intentness on substantiating its claims with evidence and its preoccupation with representing convincingly, the *Dream* also manifests a fascination with its own methods of fabricating a lunar world and its particular astronomy. More broadly we could say that it manifests a fascination with the way the imagination creates the unprecedented. This is communicated primarily through structural aspects of the *Dream*, and it is communicated only implicitly – not least because this fascination with fiction making to some extent subverts the *Dream*'s scientific argument by questioning the licitness of its presentation. In the following, we are first going to look at the way Kepler constructs the argument of the *Dream* rhetorically. Our primary concern, however, will then be the self-conscious attitude the *Dream* assumes toward the rhetorical construction of its argument and the way it implies a critique of its own use of imagery, or what could be called a poetics of the imagination. This issue, though not spelled out explicitly, is firmly embedded in the narrative and in the overall structure of the *Dream*, so much so that it becomes one of its most persistently and insistently pursued topics.

While the goal of the *Dream* is scientific, Kepler presents its theses in an intricately structured narrative, steeped in myth, legend and folklore as well as thinly veiled allusions to his own life.[9] Like the narrative it contains, the *Dream* as a book is itself a

8 Fernand Hallyn, "Le *Songe* de Kepler," in *Bibliothèque d'humanisme et renaissance, travaux et documents* 42 (1980), 343. Kepler elucidates this central thesis of the *Dream* in notes 54, 96 and 135 of his commentary. Cf. *Kepler's Somnium or Posthumous Work on Lunar Astronomy*, ed. and trans. Edward Rosen (Madison: University of Wisconsin Press, 1967). All references to the *Dream* are made to this edition, henceforth cited as *Dream*.

9 In her pioneering book *Voyages to the Moon*, Marjorie Nicolson observes: "No important later [imaginary] voyage will employ so fully the supernatural, yet none will be more truly 'scientific' than that 'Dream', which was the *fons et origo* of the new genre, a chief source of cosmic voyages for three centuries." (41, addition mine) Arthur Koestler, following Nicolson, calls it "the first work of science-fiction in the modern sense – as opposed to the conventional type of fantasy-utopias from Lucian to Campanella." Cf. *The Sleepwalkers: A History of Man's Changing Vision of the Universe* (London: Penguin, 1989), 421. For the *Somnium* as science-fiction, see also Donald H. Menzel, "Kepler's Place in Science Fiction," in *Vistas in Astronomy* 18 (1975), 896-904,

patchwork of textual components dating from different periods in Kepler's life. Since the structure of the *Dream* is of central importance to my argument, I suggest we first look at the structure of both the book as a whole and of the dream narrative it contains.

In its 1634 posthumous publication, the *Dream* included the brief fantastical tale containing Kepler's astronomy, geography and biology of the Moon (written in 1609 and partly based on a dissertation project in Tübingen dating back to 1593) as well as a series of explanatory notes (written between 1621 or 1622 and 1630), which are much longer than the narrative itself. In this commentary, Kepler indicates textual sources and performs various analyses of the material presented in the narrative. These range from allegory, autobiography and etymology to science, scientific methodology and others. After this longest section of the book, there is an appendix in the form of a letter by Kepler to the Jesuit mathematician Paul Guldin: "Geographical, or, if You Prefer, Selenographical Appendix"[10] (written in 1623). Here Kepler reports some more recent telescopic observations he made of the circular craters on the lunar surface. He interprets these craters as fortifications built by the rational inhabitants of the Moon and even offers a careful description of how such fortifications are built. This letter, in turn, is followed by further notes, which lay out the argumentative foundations for the fortification theory. The book closes with Kepler's own translation into Latin of Plutarch's *De facie in orbe lunae*, which Kepler had especially written to be published together with his lunar astronomy, and a series of notes upon that work (finished shortly before 1629).[11]

Within the narrative itself matters are no less puzzling. I will summarize here the main lines of the narrative and mention in context those aspects of the *Dream* that will be important to my argument. To avoid incessant interruption, some of my comments are placed in the footnotes. The *Dream* consists of a series of interlocking narratives, the innermost of which contains the actual lunar astronomy as well as the description of lunar topography and forms of life. The overall narrative transcribes a dream of the first narrator (Kepler), which was evoked both by reading a story about the legendary Bohemian ruler Libussa and by the narrator's observations of the Moon and stars. The dream features a book in which a second narrator by the name of Duracotus relates his fortunes – in several respects strongly reminiscent of Kepler's own.[12] The first narrator thus not only transcribes his dream, but also a book by someone else, the contents of which reflect back on the first narrator. Accordingly, the stratifications of the narrative start to blur from the moment they are introduced.[13]

---

and Gale E. Christianson, "Kepler's *Somnium*: Science Fiction and the Renaissance Scientist," in *Science Fiction Studies* 4 (1977): 79-90.

10 "Appendix Geographica, seu mavis, Selenographica". Cf. *Gesammelte Werke*, ed. W. von Dyck and M. Caspar *et al.* (Munich: Beck, 1937-), vol. 11, pt. 2, 368. All references to Kepler's writings are made to this edition, henceforth cited as *GW*. Unless otherwise noted, references refer to this volume.

11 For the estimated dates of composition, see Edward Rosen's annotations to the *Somnium*.

12 For a detailed account of the parallelisms between the story of Duracotus and Kepler's own life, see Marjorie Nicolson, *Voyages*, as well as part IV of the introduction to *Kepler's Dream*, ed. John Lear, trans. Patricia Frueh Kirkwood (Berkeley: University of California Press, 1965), 21-26.

13 In his analysis of the *Dream*'s narrative structure, Fernand Hallyn notes the symmetric inversion of relation: "The *Dream* is a book that tells of a dream. But the dream tells of a book. ... A further indication, if one were needed, that the multiplication of narrative levels only creates an apparent

Duracotus's mother, Fiolxhilda, is a wise but irascible woman, who in a blaze of anger impulsively pledges her son to the captain of a ship. After a turbulent voyage, Duracotus eventually ends up on the island of Hven, where Tycho Brahe – Kepler's teacher and employer – is conducting his astronomical observations. For some years Duracotus remains with the great astronomer, who introduces him to the science of astronomy. Tycho's observation of the stars and the Moon reminds the boy of his mother, "because she, too, used to commune with the Moon constantly."[14] (13) After five years, Duracotus returns home and is reunited with his mother. To his surprise, he comes to realize that the 'wise woman' is as wise as Tycho Brahe in lore of the skies. In some way, his mother has learned by experience all that Tycho surmised.[15] She tells Duracotus that she communicates with the "daemons of Levania" (the spirits of the Moon), with whom mortals at times travel to the distant land. In the company of her son, Fiolxhilda invokes one of the spirits while they both cover their heads with their clothes, in accordance with the magic covenant.

Here ends the first section of the *Dream*. The second, longer section consists of the Daemon's description, in Icelandic, of the journey to 'Levania' (the Moon), and of Levania itself. In addition to being a transcription of a dream and a book, the *Dream* thus also includes a translation of a narrative the book contains. Although this narrative is marked as the Daemon's, the same ambiguity exists between the Daemon and Duracotus as between Duracotus and the first narrator: the Daemon explicitly assumes a human perspective.[16] This ambiguity is emphasized in several ways: by the point about the daemons' sometimes escorting human beings to Levania, their preference for dried-up old hags, adept in the lore of daemonology (thus recalling Fiolxhilda),[17] as

---

discontinuity. From one level to the next, it is always the same that is signified." ("*Songe*," 334, my translation) See also the adaptation of this article in chapter 11 of his *Poetic Structure*.

14 *Dream*, 13. – "quippè et ipsa assiduè cum Luna solita erat colloqui." (*GW*, 322)

15 Timothy Reiss has observed that "the relationship between Duracotus and his mother, Fiolxhilda, matches that of the 'original' narrator with Libussa." (*Discourse,* 150) Thus, the parallelisms between the two narrative levels are further developed.

16 This has been pointed out by Hallyn in "*Songe*," 335, as well as by Reiss; yet Reiss makes the stronger argument that the Daemon is "metamorphosed" into Duracotus through a syntactic change from the third person initially used to refer to the Daemon to the first person. Here I do not agree, however; the Daemon's is a first-person narrative, as is clearly marked in the text with the words "It began at once as follows" as well as a title heading that sets the Daemon's narrative off from the preceding narration. Moreover, the Daemon's voice is characterized as "rasping" and "unclear" ("blaesae et obtusae vocis", *GW*, 323). This is why I think that the situation remains ambiguous. (Rosen's use of the neutral personal pronoun in '*it* began as follows' is strange, especially since at the end of the dream narrative he refers to the Daemon as 'she'. For this additional source of ambiguity, cf. note 69, *infra.*) Reiss also makes the important observation that Duracotus does not physically travel to the Moon. Yet, in this respect, too, there is some ambiguity because of the detailed description of a journey to the Moon undertaken by human beings. As I am going to show later, ambiguity is an essential characteristic of the *Dream*. (*Discourse,* 150 and 158, and *Dream*, 14-15)

17 Alongside the ambiguous identities of Kepler, Duracotus and the Daemon, there is a parallel structure of ambiguity between the "virago" Libussa, Fiolxhilda and the old witches that blurs the distinction between the three narrative levels. John Lear has observed that in Kepler's day the word 'virago' (*GW,* 321), which Kepler used as an attribute for the legendary Bohemian ruler Libussa, could have both the meaning of "a termagant, a shrew, a noisy and unpleasant woman" as well as the older one of "a female of heroic attributes." "If this old definition is accepted, it can be

well as the detailed account of the effects the trying journey has on the human body. Moreover, the description of the trip to the Moon calls to mind Duracotus' nauseating journey to the island of Hven earlier in the narrative. The fact that the Daemon refers to the Moon as "the island of Levania" and compares the arrival on the Moon to "disembarking from a ship and going ashore" further underscores this parallelism.[18] In a continuous play of parallelisms and ambiguities the various narrative levels keep implicitly referring to each other and narrative and descriptive elements are recycled from one level to the next.

In this third narrative the focus becomes primarily scientific; albeit not exclusively so, for there are still elements of myth and superstition, such as the piece of Pythagorean daemonology when the Daemon explains that he and his kindred spirits can only embark on their journey during a lunar or solar eclipse, when the cone of shadow (*umbra*) between the Earth and the Moon serves them as a ladder.[19] As soon as the daemons arrive on the Moon, they "quickly withdraw into caves and dark places" to escape from the Sun.[20] In a further narrative parallel, the daemons' flight from the Sun and the repeated references to shadow and dark places prefigure the plight of the lunar inhabitants, who spend half their lives frantically escaping from the scorching Sun. The ambiguity between the identity of Duracotus and the Daemon thus extends to the lunar inhabitants. In consequence, the demarcation of different perspectives – established on the narrative level through the separation of narrative voices, and (as I laid out in the introduction to this chapter) of central importance to the cosmological argument of the *Dream* – becomes uncertain.

The Daemon then gives a systematic description of Levania. The Moon is divided into the Privolvan and the Subvolvan hemisphere. While the Subvolvan always faces the Volva (the lunar name for the Earth), the Privolvan remains deprived of this sight. The Daemon systematically describes various phenomena of lunar astronomy and points out the differences between the two hemispheres: the Privolvan is mainly characterized by the complete absence of the Volva. This absence, as well as Levania's proximity to the Sun during the Privolvan day, generates an extreme shift in temperature between day and night. The Subvolvan, on the other hand, enjoys the great spectacle of the Volva, which replaces our Moon. Its waxing and waning as well as the way it turns on its axis are described very beautifully and in great detail.

Finally, the Daemon describes Levania itself. Topographically, the lunar world is much like ours, save that everything is bigger in scale: the mountains soar to incredible heights, the fissures and valleys are deeper than any terrestrial canyon. Despite Gali-

---

the older one of "a female of heroic attributes." "If this old definition is accepted, it can be seen that Kepler once again was punning when he wrote of Libussa's 'magic art.' ... Kepler would have been justified in making one of his typical quips about her extraordinary powers, for her gift of prophecy was such as to bring her the nickname, Sibylla Bohemica." *Kepler's Dream*, 74-75. Accordingly, all the women who make their appearance in the *Dream* are simultaneously wise women and witches.

18 *Dream*, 15-16. – "LEVANIA insula"; "quasi ex navi in terram exscendentes". (*GW*, 323-24) On this parallel, see also Reiss, *Discourse*, who points out that the similarity is also emphasized in the notes. (157)

19 For Kepler's use of this and other folkloric elements, see Hallyn's argument summarized in note 98, *infra*.

20 *Dream*, 16. – "et ibi nos properè in speluncas et loca tenebrosa recipimus". (*GW*, 324)

leo's growing suspicion to the contrary, Kepler believed in the existence of water on the Moon and therefore in the existence of human life. Forms of life on Levania have adapted to the topographical environment and the great calamities of the climate. Creatures and plants are of monstrous size,[21] but their life span is brief and seldom lasts more than one lunar day. Of the two hemispheres, life on the Subvolvan is more bearable for the presence of the Volva and the Moon's aphelion position during the Subvolvan day.[22] In addition, a constant cover of clouds and rain give relief from the heat. From this vision, the narrator is suddenly roused by the rattle of rain to find that it was all a dream, that he is lying in bed his "head really covered with the pillow and [his] body with the blankets."[23]

Three elements receive special prominence in the narrative of the *Dream* and are continually played off against each other: the book, the dream and the journey. The three elements mutually contradict each other: The physical journey to the Moon conflicts with the dream vision, and this in turn with the dream as book or literary dream. They can only exist next to each other through ambiguity. At the same time, each of them relates to the contents of the *Dream* in a particular way: It is either constitutive of its argument or it reflects on its mode and validity of representation through critical distancing. My discussion will focus mainly on the concepts the three elements symbolize in the *Dream* and their relation to each other. In the combination of these elements with both the structure of the narrative and the structure of the *Dream* as book, the text manifests a self-consciousness about the way it presents its argument, most evident in the way the *Dream* playfully oscillates between book, dream and journey. Since we have just recapitulated the narrative within the *Dream* and since the element of the journey is restricted to this narrative, we will start with the journey.

In the introduction to this chapter, I pointed out that the astronomical thesis of the *Dream* is based on a change in perspective that serves to counterweigh the familiar terrestrial appearance of the heavens with their appearance from the Moon. Living beings on the Moon, according to Kepler's explicit thesis in the *Dream*, would not only be likely to conceive of the cosmos as lunocentric and lunostatic, they would, moreover, witness the spectacle that to a human being on Earth was so difficult to accept: the Earth spinning around itself once every twenty-four hours. To these two arguments for the movement of the Earth – the analogy of the two boats and the spectacle of the spinning Earth – Kepler's lunar astronomy adds a more innovative and rhetorically subtle point in defense of Copernicanism. It concerns the objection often raised against Copernican theory that if the Earth did indeed revolve around the Sun, then one would have to be able to measure a parallax of the fixed stars from different positions of the Earth's orbit. Yet with the optical instruments available at the time, no parallax was measurable.

In order for Copernican theory to be correct in spite of this circumstance, the distance between the Earth and the fixed stars had to be incomparably larger than the diameter of the terrestrial orbit. As a result, the latter shrinks to a mere point in comparison. A parallax of the fixed stars would then only become visible if one compared the

---

21 As noted in the Introduction, this idea is taken from Lucian's *True History*. Cf. p. 11, *supra*.

22 The second point illustrates Kepler's theory of elliptical planetary orbits.

23 *Dream*, 29. Addition mine. – "verè caput pulvinari, corpus stragulis obvolutum deprehendi." (*GW*, 331)

position of the fixed stars as seen from the Earth with the position they occupy from a point beyond the terrestrial orbit. This condition of dimension, the idea of so much empty space between the planets and the fixed stars, was among the points that caused most opposition to the theory.[24] Kepler realized that due to the relative closeness of the Moon to the Earth, the parallax of the fixed stars would not be visible from the Moon either. Yet he reckoned that from the Moon the various movements and the sizes of the planets would appear much differently than on Earth. This is what the Daemon notes in his description of the lunar sky:

> The fixed stars look the same to all Levania as to us. ... But its view of the movements and sizes of the planets is very different from what we observe here, so that its entire system of astronomy is quite diverse.[25]
>
> Etsi siderum fixorum aspectus tota Levania habet nobiscum eosdem; motus tamen Planetarum et quantitates, ab ijs, quas nos hîc videmus, observat diversissimas: adeò ut planè alia sit totius apud ipsos Astronomiae ratio.[26]

Here Kepler is turning the tables on the opponents of Copernican theory, for it is now the very lack of a stellar parallax that is inconsistent with the Ptolemaic system: In a geocentric cosmos, such a parallax would have to be discernible from the Moon's off-centered position, unless one allowed for an infinitely larger distance between the planets and the fixed stars, which is just what Kepler aimed at. Yet despite their argumentative acuteness, neither this nor, in fact, any of the astronomical theses Kepler presented in the *Dream* could be considered irrefutable arguments for Copernicanism: what he lacked was empirical data, the actual *proof* that this is what the universe looks like from the Moon. So he simply invented it in his fictional narrative.

This is where the narrative element of the journey comes in. The fictional device of the journey accords to Kepler's description of lunar astronomy and topography their requisite evidentiary status – even while the Kepler of the notes smiles roguishly at this hoax.[27] Kepler presents the astronomical thesis of the *Dream* as travel literature, the description of an actual experience. The careful description of the journey to the Moon as undertaken by human beings helps to emphasize this, as does the ambiguity evoked through the parallelism between Fiolxhilda and the old witches who travel to Levania.[28]

---

24 Timothy Ferris has pointed out that besides the argument of the futility of so much empty space, "[a]ll geocentric, immobile-earth cosmologies tended to inhibit appreciation of the true dimensions of space" for the simple reason that, in a geocentric cosmos, the larger the spheres of the planets and stars, the faster their rotating speed would have to be. Cf. *Coming of Age in the Milky Way* (New York: Doubleday, 1988), 34. Adaptation mine.

25 *Dream*, 17.

26 *GW*, 324.

27 Kepler repeatedly refers to the dream narrative as a 'joke'; cf. notes 41, 59, 61 and 86.

28 "We especially like dried-up old women, experienced from an early age in riding he-goats at night or forked sticks or threadbare cloaks, and in traversing immense expanses of the earth." (12) Fiolxhilda, moreover, makes "little bags out of goatskin" with magical power. (15) For the ambiguity between Libussa, Fiolxhilda and the witches, see note 17, *supra*.

Besides the fictional device of the journey within the narrative itself, the impression of an actual experience is conveyed by the *Dream*'s peculiarly divided structure.[29] It is remarkable how all kinds of explanatory comments are strictly absent from the narrative. The world in the Moon is, as it were, simply described as it manifests itself to someone on the Moon; all the explanations and comments as to what induced Kepler to describe lunar astronomy and the Moon itself the way he did only appear in the notes. In this connection a passage from a letter Kepler wrote to his friend Matthias Bernegger is instructive. There, Kepler presents the dream narrative as a kind of riddle for astute readers:

> My treatise contains as many problems as lines, which may be solved partly with the help of astronomy, partly with that of physics and history. Yet who will think it worth the trouble to solve them? People want that such a plaything be presented to them in a leisurely way; they do not like to knit their brow while they are playing – which is why I decided to resolve everything in notes following the text.
>
> In mea commentatione totidem sunt problemata, quot lineae, solvenda partim Astronomicè, partim physicè, partim historicè. Sed quid agas? Quotus quisque curâ dignum censebit, ea solvere? Cupiunt homines, hujusmodi ludicra sibi molli brachio, quod ajunt, obtrudi, nec faciles sunt ad corrugandam frontem in ludo. Itaque statui notis post textum, numerorum ordine sequentibus omnia solvere.[30]

Is Kepler just being over-modest when he claims that in offering readers a commentary to his lunar astronomy, he is merely sparing them the trouble of deducing the implications of this astronomy for themselves? Surely he must have realized just as much as Bernegger that an important part of his achievement in the *Dream* lies in explaining how the phenomena he describes are produced. This is expounded in all its consequences only in the notes. Kepler could not seriously have expected even his most discerning readers to be able to read his little dream narrative and from it deduce not just the cosmology, but the physical and optical laws responsible for the phenomena described in the narrative. While expectations of this kind, though still unlikely, may be conceivable with regard to phenomena based on anything ranging from received theory to ongoing astronomical debates, they must be ruled out when it comes to the scientifically innovative parts of the *Dream*.

In his introduction to the first published English translation of the *Somnium*, John Lear notes two examples: the first is Kepler's pioneering contribution to the development of the principle of inertia – what could be called the inertia of rest, to which Newton was to add that of motion. The *Dream* represents a vital step in the development of this principle. The other example is Kepler's theory of tides; it attributes the phenomenon to the reciprocal gravitational pull, which the attractive forces of Moon,

---

29 With the notable exception of Timothy Reiss, whose reading of the *Dream* and its structure will be referred to in more detail later on, this characteristic of the *Dream* has received little attention. Fernand Hallyn points out in passing that "[t]he principal literary example of a work in which the author's annotations surpass the length of the text itself is Guarini's *Pastor fido*." He argues that the disproportionate relation between text and annotations reveals "a self-engrossed attitude toward writing, [a] generation and over-determination of meaning." (*Poetic Structure*, 255)

30 Letter of 4 December 1623, *GW*, vol. 18, 143. German translation mine.

Earth and Sun exert on the waters of the Earth and the Moon.[31] In the *Dream*, Kepler no longer restricted mutual gravitational attraction to kindred bodies (as he had done in his *New Astronomy*), but more generally applies this force to different bodies. Here Kepler took a long stride in the direction of Newton's universal gravity.[32]

The narrative dramatizes both the principle of inertia and the concept of mutually attractive forces between heavenly bodies, but it is only in the notes that they are set into an explanatory framework.[33] The description of lunar astronomy and the world in the Moon thus serve as illustrations of the theoretical concepts elaborated in the notes, but they are not self-explanatory. Rather, they represent the sensory experience that the commentary analyzes and interprets.

Through this structural division of the text, the dream narrative comes to represent the empirical ground on which Kepler bases his theories. By offering his description of a lunar astronomy and world in a separate narrative, he presents it as pure and undefiled by deductive and hypothetical reasoning. The reader is first confronted with the description of phenomena and is then given an explanation for them. Kepler thus reverses the direction of the actual deductive process, obscuring the fact that he fabricated his description primarily on the basis of a cosmological hypothesis: The lunar astronomy described in the narrative, so it appears, is where the deductive process starts out, not where it ends. Moreover, by professing that the solution he supplies in the commentary is not requisite to understanding the implications of the lunar astronomy, but only facilitates the reading, Kepler suggests that the solution is naturally entailed by the problem, that the judicious reader will automatically derive the correct cosmological conclusion from this lunar astronomy – just as he himself did from the corresponding terrestrial astronomy. What the rhetoric covers up is that this conclusion will only be drawn spontaneously if read in conjunction with the corresponding terrestrial astronomy and that the lunar astronomy remains hypothetical and thus a fiction.

And yet this is only partly true; for, as Kepler asserts in his letter, the narrative is a 'plaything', and a many-faceted one. Since Kepler is inviting his readers to join him in playing, let us look more closely at this plaything and the kind of tricks it performs. The narrative does not just relate the experience of a journey to the Moon; what it relates is a dream of a book which relates a journey to the Moon or, more precisely, a dream of a book in which someone relates how he was listening to someone else relating the experience of such a journey. At the same time, we are offered not just this dream experience, but its retrospective presentation in a systematically ordered narrative, where the dream is set into the context that allegedly produced it.

The narrative thus plays with the forms and identities of its contents. It is about representation just as much as about experience or, as we shall see, experience in the

---

[31] *Kepler's Dream*, 64 and 71-72.

[32] Edward Rosen's discussion in Appendix H to his edition of the *Dream*, 219.

[33] In the narrative, both the principle of inertia and of mutual gravitational attraction are dramatized in the description of the effects of changing gravitational fields on the human body during its passage to the Moon and the reference to a field of zero gravitational force. An illustration of Kepler's theory of tides is given through the vivid account of torrential tides on the Subvolvan hemisphere, which are attributed to the conjunction of Sun and Earth pulling all the water on the Moon in their direction. Cf. pp. 16, 18 and 27. Kepler's explanation for the first two principles can be found in notes 66, 67, and 74 through 76. His theory of the ebb and flow of tide in notes 62 and 202 through 204.

imagination. It is about the indirection and the artifice of representation, about the power of evoking a scene in someone else's mind through description and about envisioning something in one's own mind, about the immediacy of what we imagine – all at the same time. The narrative shows simultaneously what is imagined and the act of imagining and representing it: it is at once what Kepler sees in his dream, what Duracotus imagines while listening to the Daemon's description of Levania, and what the Daemon himself sees in an actual experience; but it is also the Daemon's *narrative* of this experience aimed at evoking a particular scene in the imagination of his two listeners, Duracotus' own rendition of this narrative aimed at evoking the same scene in the imagination of the readers of his book, and Kepler's reproduction of both of these narratives in his dream, this time aimed at the readers of his own book, that is, at us. The narrative thus not only evokes a particular scene in our imagination, it also shows how we, as readers, are being manipulated into imagining something.

There is a constant shift of focus between the two conflicting concepts of experience and representation. Speaking in terms of the three narrative elements I pointed out earlier, these two concepts are most plainly symbolized by the elements of the journey and the book, respectively. However, the Daemon's physical journey to the Moon, which forms the centerpiece of the narrative, is at least partly superseded by the act of *imagining* the astronomy and the topography of the Moon *as an actual experience*, which Duracotus, Kepler and we as readers perform on the different levels of reception. While, as we have seen, the theme of the physical journey to the Moon is of fundamental importance to the (fictional) validity of the argument the *Dream* presents, the act of imagining the experience of such a journey plays an equally important role in the rhetoric of the *Dream*, on account of its persuasive power. Kepler helps his readers visualize a lunar astronomy and world as vividly as possible, and in this way tries to make his theory palatable to his audience. At the same time, this experience in the imagination, which factually invalidates the description of a journey undertaken in person, owes much of its sense of realness to the narrative frame of the journey, since it is the fictional device of the journey that provides the setting for an experience of vivid immediacy in the first place. This sense of realness is an effect of the sharp division between description and commentary in the *Dream*'s structure; by keeping all explanation and methodology out of the Daemon's description, Kepler creates a sensory experience in which he invites his readers to participate in their imagination. Thus, actual experience is superseded by experience in the imagination, a concept most adequately symbolized by the third narrative element, the dream.

It is significant that the mental act of imagining an experience should be rendered through the symbol of the dream, for this symbol encompasses experience and representation as well as imagination, i.e., experience and representation in the mind. Kepler's *Dream* uses all three of these and plays them off against each other. As a symbol, the dream at once illustrates the argumentative framework of Kepler's text in its central characteristics *and* causes it to reflect on itself as an artificial creation.

This is suggested in the triple ambiguity between the three narrators: the Kepler who is dreaming as well as narrating is at the same time the Duracotus who is imagining as well as narrating and the Daemon who is narrating an experience. The experience of dreaming is at once real and fictitious; that is to say, dreaming resembles actual experience in that what we experience when we dream appears with the vivacity of an actual experience rather than something we consciously imagine. At the same time, dreams relate to the imagination in so far as they, too, are only chimeras of the mind, produced on the basis of past experiences. It was this characteristic that induced Aristotle to classify dreams as a subcategory of imagination, or what he called *phantasia*;[34] and the same characteristic lies behind Kepler's parallel treatment of dreaming and imagining.

Dreaming could thus be compared to a particularly vivid, unconsciously performed act of the imagination. The juxtaposition of Kepler's dream with both Duracotus evoking in the imagination what the Daemon relates and the readers of the *Dream* doing the same with Kepler's relation on an implicit level outside the narrative in fact clearly suggests this kind of relationship. Finally, Kepler makes it very clear that his dream narrative is a written text and thus not a real dream, but an imitation of a dream; as such it belongs to a well-known literary genre.[35]

Within the narrative itself, the *Dream*'s identity as narrative and text is made prominent both through the stratification into different levels of narration and the narrative element of the book, which appears on two of the narrative levels. The multiplication of the act of representing draws the reader's attention to the fact that the experience which the *Dream* presents is first and foremost a narrative construct. The book at large emphasizes the same aspect, first of all, through the variety of the book's textual components, that is, through the way the dream narrative is, as it were, embedded in text and only represents one component of a larger work which Kepler composed over a long period of time. Yet, more particularly, it is the commentary and the way the commentary interacts with the narrative which lends prominence to the textuality of the dream narrative and characterizes it as a literary creation.

The stratification of narrative levels within the narrative and the interaction of narrative and commentary in the book also focus on another aspect – one which, as the genre of the literary dream suggests, directly relates to the aspects of composition and the *Dream*'s textuality: representation in the imagination. A major concern in the *Dream* lies in documenting the coming-into-being of its dream narrative and the description of the lunar world as a creation of the imagination. While the commentary offers the documentation of this creative process, this documentation goes hand in hand with a paradigmatic representation of the formation of dreams within the dream narrative. The parallelism between dreaming and imagining is thus once more central; it now includes a third process, that of literary composition. Finally, the *Dream* also addresses the question as to how – given the conditions of its production – this product of human fantasy may be considered a representation of another world that both in its potential otherness and the actuality of its existence is beyond the reach of human imagination.

The way the *Dream* presents itself remains ambivalent throughout. As the structure of the dream narrative shows, there is a constant tension between the drift of the

---

[34] Cf. p. 5, *supra*.

[35] For the *Dream* as written text, see also Hallyn, *The Poetic Structure of the World*, 257. However, I do not agree with Hallyn when he argues that the *Dream* should be considered *exclusively* as "a *written text*, for which the appropriate operation is not reconstituting an experience but analyzing the meaning."As an imitation of a dream, Kepler's narrative is also an imitation of an experience, and it is one of my main concerns here to show how Kepler exploits this ambiguity between felt and crafted experience.

Dream's rhetoric, which fashions the dream narrative as a description of an authentic experience, and the implications of its self-questioning. On the one hand, the narrative stratification of the text directs the reader's attention to questions of representation and textuality, thus questioning the authenticity of the experience; on the other hand, the ambiguity between the narrative levels weakens this effect and thus to some degree restores a sense of unity within the narrative and with it the impression of a singular experience.[36]

The same kind of tension runs through the whole book, though working in the opposite direction: I mentioned earlier that the sense of unity and immediacy of the experience described in the narrative is to a large extent achieved through the neat separation between description and commentary. The impression of a unified experience is, however, disturbed by the numbers referring to the commentary, which constantly interrupt the flow of reading and with it the readers' effort to duplicate in their imagination the experience described. This sense of duplicity is a constant presence in the *Dream*, and it remains – by necessity – unresolved: the text keeps focusing on its own rhetorical mechanism even while it is intent on achieving its most powerful effects. It is, as it were, dreaming itself and exposing itself as a mere dreamer.

The illusion of an authentic experience (be it actual, imagined, or dreamed) which the fiction creates is exposed by the commentary as a fabrication right from the start. Here Kepler, who in the fiction figured as the passive dreamer, placed at third remove from any immediate participation, steps forward as author of the text and assumes authority over its composition. And whereas as readers of the narrative we were made to participate in the experience, we are now asked to witness the compositional process of the dream narrative both as a text and a creation of the imagination, and to evaluate as scrupulously as possible the many considerations that went into its making.

The commentary, which in length surpasses the dream narrative several times over, exhibits an almost obsessive concern with documenting the genesis of the narrative. This is not only the case with respect to the astronomical and selenographical part of the text, where the notes provide the scientific basis for the phenomena described by linking them to observations and mathematical calculations, but just as much – and more obtrusively – with the framing narrative. There is no element that goes uncommented: Kepler explains the choice of names for characters by pointing out etymological derivations, and the choice of setting by listing geographical particularities which suit the context; he attributes allegorical significations to the various characters and relates narrative episodes to popular myths, literary precedent, autobiographical experiences and historical events. The commentary shows us Kepler's imagination at work in as much as it indicates the source of inspiration for basically every single narrative element and records the story of how each element found its way into the text. It also shows how, as a result of new ideas Kepler got by reading other fictions, the plan of what was to become the dream narrative changed over time and how such ideas at times resulted from pure chance.[37] Since this is done in the form of notes that interrupt

---

36 See also the point made by Hallyn, quoted in note 13, *supra*, that the multiplication of narrative levels creates only an apparent discontinuity.

37 A good example for this may be found in note 2, where Kepler first relates how he read Lucian's *True Story* to master his knowledge of Greek and came across a gripping account of a voyage to the Moon: "He [i.e., Lucian, as author], too, sails out past the Pillars of Heracles into the ocean

the text, it is in the course of reading the narrative that we become aware of the process of textual composition and of the way the text integrates material from an abundance of other texts and sources.

Besides this preoccupation with the fable's derivation, the commentary is also more generally marked by a fascination with possibilities of faking, with the crafting of illusions and myth-making. This becomes evident at the points where the commentary seems to be at its most obsessive with regard to what it undertakes to explain. In note number 50, for instance, Kepler goes about explaining how the apparently supernatural voice of the Daemon could have been produced mechanically with some kind of talking machine; he adds that "in this mechanism ... there are built-in traps for the superstitious and gullible, so that they may sometimes suppose demons are talking to them when art is copying magical tricks."[38] Rather than as part of a fictitious story, the comment refers to hearing the Daemon's voice as an actual experience – though an illusory one. And yet the question it attempts to solve is how this particular experience could be faked with human means, as a special effect, so to speak.[39] Note number 15, which refers to Fiolxhilda's earning a living by selling bags of goatskin to ship captains, works similarly: It first of all ties this passage of the dream narrative to its inspirational source by referring to a popular saying which holds "that the pilots of ships sailing from Iceland produce whatever wind they want by opening a wind bag." The comment first integrates the passage into a popular myth; it then deconstructs this myth, spelling it out in terms of what it might realistically refer to: the expert helmsman who employs "the rhumbs of the nautical rose, magnetic needle, and rudder control" in such a skilful way as to steer his ship in any desired direction – *as if* by supernatural means.[40]

Both of these comments reveal a special concern with the relation between the fictional and the factual or realistic and with the question of the factual basis of fictions: by pointing to possible mechanical origins behind a seemingly supernatural experience, the first comment considers possibilities of faking an experience and thereby implies

---

and, carried aloft with his ship by whirlwinds, is transported to the moon. These were my first traces of a trip to the moon, which was my aspiration in later times." Explanation mine. In the same note, Kepler explains that his decision to set the *Dream* in Iceland originated from some stories that chanced to be bound up with a German translation of Lucian's book he acquired at a later date. One of these stories was about St. Patrick's purgatory, purportedly located "in the earth beneath Mt. Hekla, the Icelandic volcano." Besides choosing Iceland as a narrative setting for his tale, Kepler also refers to Mt. Hekla. The element of the purgatory is fused with a passage in Plutarch's *De facie*, where the Moon is referred to as the purgatory of souls, to produce the daemons traveling back and forth between Mt. Hekla and the Moon.

38 "Et in hac mechanica puto nonnullas esse collocatas insidias superstitiosis et credulis, ut quandoque existiment, sibi loqui daemonas, cùm praestigias magicas ars imitetur."

39 In notes 44, 46 and 47, Kepler compares the ceremony around the Daemon's apparition with the ceremony he used to perform with spectators witnessing observations and experiments he conducted at his house in Prague. Interestingly, it is Kepler's alter ego Duracotus who is fooled in the narrative rather than being the one who fools. This seems to suggest that in the *Dream* Kepler is not the magician, but is himself an enchanted spectator who is taken in by the performance, even though he knows it to have been produced artificially.

40 "Gubernatores navium ex Islandia navigantes, Ventorum quem velint, aperto Venti utre excire"; "quod si quis ad Rosae nauticae Rhombos, indiculumque magneticum, et rectionem gubernaculi transferat, is vera propemodum dixerit."

that the entire dream narrative is a mere fabrication; the second traces the possible factual origins of a myth and thus suggests how a myth comes into being.[41] Both focus on the issue of production and genesis.

The comment regarding the wind bags can, of course, also be considered the other way round, i.e., from an interpretive rather than productive perspective: Not only does it demonstrate how a popular myth may be traced back to factual origins, it also proposes an interpretation of the myth that approximates its contents to a more realistic setting. Here as well as elsewhere the commentary shows that the narrative requires translation and interpretation.

All of this suggests that we should not take anything presented in the *Dream* simply for what it is, but should always ask about the provenance of each narrative component and its possible signification. The commentary insists on making it clear that what we read is neither an original nor a literal representation with respect to the terms in which it is presented. It is not original, because each of its elements is either borrowed or adapted from another source, and it is not literal, because what it attempts to represent is rendered in a strange guise. What kind of representation is it then?

Fernand Hallyn has shown that the answer to this question is contained in the work's title, which implies that the representation is allegorical. Let us briefly recapitulate his argument: He points out that while the term *somnium* could apply to dreams in general, in a more technical language – which Kepler, as an avid reader of treatises of oneiromancy, was evidently familiar with – it designated a specific kind of dream that stood in contrast to *vision*. Macrobius was one of the principal authorities on the matter. In his *Commentary on the Dream of Scipio*, a work to which Kepler often referred, he distinguished five main types of dreams and defined *somnium* as a dream which "conceals with strange shapes and veils with ambiguity the true meaning of the information offered, and requires an interpretation for its understanding." Cicero's *Somnium Scipionis* is a literary imitation of this kind of dream. As such, Macrobius has classified it "among 'fables' that it is appropriate for the philosopher to use. ... Such a dream 'rests on a solid foundation of truth, which is treated in a fictitious style.' The content is always allegorical."[42]

Cicero's work is already evoked in the title of Kepler's *Dream*, where *Somnium Scipionis* is reflected in *Ioh. Keppleri mathematici olim imperatorii somnium*. In listing the literary models of his *Dream*, Kepler moreover refers to Cicero's *Somnium Scipionis* first and foremost. When he explains his choice of Iceland as Duracotus's native place, Kepler observes:

> But in this remote island I perceived a place where I might fall asleep and dream, in imitation of the philosophers in this branch of literature. For Cicero crossed

---

41 In the narrative by Sulla which ends Plutarch's *De facie*, there is a parallel passage where the classical myth of Persephone's wanderings is said to "contain the truth [spoken covertly]." Accordingly, the myth is to be read as an allegorical account of the recurring phenomenon of the lunar eclipse. However, as Harold Cherniss notes in the commentary to his edition of the text, this reading requires some major changes in the mythical narrative. Cf. pp. 193-95. Emendation by Cherniss. I refer to this episode here because it represents only one of many striking parallelisms between Kepler's narrative and the one by Sulla that reflect on questions of representation. We will turn to the others at the very end of this chapter.

42 *Poetic Structure*, 256-57.

over into Africa when he was getting ready to dream. Moreover, in the same western ocean Plato fashioned Atlantis, whence he summoned imaginary aids to military valor. Finally, Plutarch, too, in his little book on *The Face of the Moon*, after prolonged discussion ventures out into the American ocean and describes to us such an arrangement of islands as a modern geographer would probably apply to the Azores, Greenland, and the territory of Labrador, regions situated around Iceland.[43]

In hac verò remota insula locum ego mihi dispexi dormiendi et somniandi; ut imitarer Philosophos in hoc genere scriptionis. Nam et Cicero trajecit in Africam somniaturus, et Plato Atlanticam in eodem Oceano Hesperio fabricatus est, unde fabulosa virtuti militari subsidia accerseret; et Plutarchus denique, libello de facie Lunae, post multum sermonem in Oceanum Americanum exspaciatur, describitque nobis situm talem insularum, quem Geographus aliquis modernus Azoribus, et Gronlandiae et Terrae Laboratoris regionibus circum Islandiam sitis probabiliter applicaverit.

By pointing out that his *Dream* is written in imitation of Cicero's *Somnium Scipionis*, Kepler presents his own narrative as an imitation of an imitation of a dream – and thus, implicitly at least, not just as an allegory, but rather as an imitation of an allegorical representation. We will consider in a moment what this might mean, yet let us look first at instances where the dream narrative is explicitly characterized as an allegory.

There are several places in the commentary where Kepler suggests that we read his dream narrative as an allegory, especially the framing narrative around the description of a lunar astronomy and topography. He even offers possibilities of allegorical interpretation. Accordingly, Duracotus could stand for Science, his mother for Ignorance or purely practical experience. The father, who died very old when Duracotus was still an infant, could signify Reason; he remains anonymous because Ignorance cannot know the identity of Reason. Moreover, the fact that Duracotus can only start to write after his mother's death shows that not until Ignorance has been dissipated can Science reveal to the public the deeply hidden causes of things.[44] The daemons – their name, Kepler explains, derives from the Greek verb *daiein*, 'to know'[45] – represent particular sciences; the specific daemon that appears to Duracotus and Fiolxhilda represents astronomy. If we read it in this light, "the passage through the *umbra*, of Pythagorean origin, takes on a new sense: 'If we continue the allegory, [we must understand that] it is easy for reason with the help of shadow measurement [i.e., measurement of the *umbra*] to attain knowledge of celestial phenomena'."[46]

---

[43] *Dream*, note 2. I am quoting Kepler's note at more length than may seem necessary here to include his reference to Plutarch's *De facie*. As we will see, the fact that Kepler specifies Plutarch as a source is also significant in the context of dreams since the particular passage referred to by Kepler features another dream of a particular sort.

[44] Cf. notes 4 and 10.

[45] Cf. note 34.

[46] Hallyn, *Poetic Structure*, 267-68. Explanations to passages quoted from the *Somnium* have been added by Hallyn. Timothy Reiss has pointed out to me in correspondence that Kepler's allegoric key to his dream narrative may present more than a memory of the *Tabula of Cebes*, a moral allegory containing a mixture of Socratic, Stoic, Pythagorean and Cynic elements and probably the most widely used textbook of the sixteenth century. Having looked at the text myself, I think that Kepler must almost certainly have used it as a source: The *Tabula* narrates an allegory of the hu-

Kepler is suggesting here that we read the tale involving Duracotus, his parents and the Daemon as a representation of Kepler's struggles to convince the authorities with his argument in favor of the motion of the earth. This kind of allegorical interpretation is hardly permissible, however. Timothy Reiss has pointed out, in fact, that the entire genre of the allegorical dream is being undermined in the *Dream*, for the type of discourse the allegorical dream represents

> could certainly not be reduced as it is by Kepler to the production of a discourse by and about a self (I am thinking of the autobiographical notes and the *I* of the notes) whose intention is given as providing truths about the world. The allegorical interpretation is explained in the *Somnium* as a merely convenient way to pass from the expression of self to the presentation of the scientific material.
>
> There is no mystery to which allegory can offer an initial approach: on the contrary, allegory is no more than a discursive device connecting the thought of the self (and its expression) to the grasping of referential truth.[47]

If the fact that the interpretation does not comply with the generic requirements of the allegorical dream raises doubts about the motivation behind the allegory, its function is further called into question by other factors: For one thing, the role the interpretation assigns here to the mother does not in the least tally with her character in the narrative, where Duracotus learns about heavenly phenomena – i.e., the allegorical signification of the Daemon – through his mother's mediation. The interpretation Kepler proposes in the commentary therefore does not fully encompass the literal narrative, as would be required of a satisfactory allegorical reading.

Contrary to Reiss' opinion that Kepler's text undermines the entire genre of the allegorical dream, Fernand Hallyn considers the allegory of science which Kepler proposes in the commentary to be "completely classical". He nonetheless points to another central formal requirement of allegorical interpretations that the *Dream* does not meet:

> To be fully achieved, this kind of reading must replace the fiction with another discourse that sticks to its own field of interest, while coherently utilizing all the

man path through life, but above all of the path to true knowledge, which is acquired through true education. Deceit (*apate*) is sitting on a throne next to the entrance to life and always threatens to lead the travelers astray by inculcating error and ignorance into them as they pass by her. The principal aim of the journey is to overcome the amount of error and ignorance swallowed at the beginning. For most people, the search for true education (*paideia*) leads through false education (*pseudo paideia*), which though it must eventually be abandoned, may be useful to arrive at true education more quickly. False education consists of the seven liberal arts (most prominently poetry) and different philosophical schools ("Hedonists, Peripatetics, and all others like them," 83). Although, according to the *Tabula*, true education is first of all a moral education aimed at the pursuit of virtue, the idea that traditional education, if viewed with a critical eye, may prove useful as a first step toward true education is clearly evoked in Kepler's reinterpretation of old myths and fables in the service of his new theories; it is also taken up on the fictional level in the way Fiolxhilda's teaching (Ignorance or purely practical experience) must eventually be overcome in favor of true science. Finally, another clear precedent to Kepler's dream narrative is the *Daimon*, whose task it is to send the travelers on their journey through life and command them to be confident. Cf. *Tabula of Cebes*, Greek and English edition, ed. and trans. John T. Fitzgerald and L. Michael White (Chico, CA: Scholars Press, 1983).

47 *Discourse,* 149.

> components of the stylized structure. The allegorical meaning shows itself to be the transcendent support, the origin and the exclusive end of the letter of the text. In the commentary on the *Dream*, however, things do not quite work out this way. ... Thus some notes express a lack of certainty about which allegorical key to adopt – a lack of certainty by which the writer distances himself from the hidden meaning. The daemon of astronomy is called by a name composed of twenty-one characters. Kepler claims he is perplexed, because this could correspond to the number of letters in *Astronomia copernicana* or the number of possible conjunctions between the planets, or even the number of possible combinations on a throw of the dice ....[48]

We are thus left with the dilemma that even while the text is openly classified as an allegory by its title, the interpretive readings Kepler offers in the notes do not provide a satisfactory and conclusive interpretation of the allegory. At odds with the argument of the narrative or vague in their own multiplicity and, as a consequence, disruptive, they are unable to get a firm grasp of the allegorical narrative; the latter maintains its independence and remains the predominant narrative, more satisfying than its interpretive translations, because more coherent in its narrativity.[49]

Are we to conclude then that it is all for nothing, or – as the reference to the throw of the dice seems to indicate – that Kepler introduced the interplay of allegory and interpretation only as playful entertainment? We will see that this is decidedly not the case and that the concept of allegory is, in fact, central for a deeper understanding of the *Dream* and the kind of claims it makes with regard to what and how it represents. Therefore, I do not agree with Reiss's claim that the genre of the allegorical dream is being undermined in this text. While it is true that the allegory which Kepler identifies in the notes and for which he offers us a rusty key fails and that, as Reiss plainly shows, it is not an allegory in the proper sense, there is a way in which the structure of allegory is basic to representation in the *Dream* as a whole – though to call this allegory is again neither classical nor entirely proper.

When the interpretations offered in the notes fail to transform the allegorical narrative in any definitive and stable way, this, first of all, calls attention to the process of translation essential to all allegory. Yet the interaction between narrative and commentary focuses on the more basic concept of allegory as a mode of representation rather than on the concrete translation of one discourse into another: a fiction that functions as a graphic substitute for another discourse that remains implied, a metaphorical nar-

---

[48] *Poetic Structure,* 268. In a footnote, Hallyn refers to another example in note 35, where Kepler comments on why "the most important daemons are nine in number. Again, Kepler gives several possibilities, but finally concedes that "the real reason escapes [him]." Adaptation mine.

[49] Hallyn's critical observations about the ambiguous allegorical interpretations which Kepler offers in the notes are very suggestive, but the phenomenon remains unexplained. In addition to drawing attention to Kepler's declared irresolution with regard to the correct interpretation of the allegory, Hallyn points out that these interpretive proposals stand side by side with notes tying the narrative to episodes in Kepler's personal past, legends and other material. He then distinguishes between, on the one hand, the allegorical reading, which leads to interpretation and tends to unify narrative units to make them homogeneous, and, on the other, comments leading to explanation, which emphasizes heterogeneity. In view of his remarks on interpretive ambiguity, I find it surprising that Hallyn should make a case for the unifying tendencies of interpretation and that he does not develop the point that in this case the interpretation is *not* homogenous. (*Ibid.*, 268-69)

rative. The failure to interpret coherently and conclusively, rather than undermining the allegory as a genre, shifts the focus to what an allegorical discourse is designed to do; it emphasizes its potential to signify something by means of something else.

Moreover, Kepler's repeated declarations of uncertainty regarding the allegorical meaning of some passages as he had originally intended it reveal the arbitrary nature of the relationship between the allegorical narrative and its underlying interpretation. They characterize allegory as a wholly man-made construct, a simple tool with no intrinsic connections to what it is used to represent. The only mystery the allegory poses must be put down to a failing memory. In documenting Kepler's struggles to find the originally intended meaning behind the allegorical narrative, the commentary testifies to Kepler's attempt to comprehend his own thought processes. In doing so, it draws attention to how the allegory was composed in the mind of its author. The main issues raised with respect to allegory are thus substitution, composition and translation.

In considering the implications of the use of allegory in the *Dream*, it must be pointed out that all reference to allegory and its interpretations is strictly limited to the framing narrative. It ends with the daemons' arrival on the Moon. Does this mean that such references are marginal? If so, then not in the sense of being *unimportant*, but rather in that of *constituting a margin or frame*, of serving as an introduction; for what else could be the function of a framing narrative if not that of introducing what is to come by setting a particular focus? In our case the focus is twofold: First of all, the fact that the allegory will not be resolved shifts the focus to allegory as a mode of expression rather than its specific interpretation; but then it also points beyond the frame and implies that the real allegory is yet to come. In other words, the narrative frame and its aborted interpretation offer a model for the rest of the text. They indicate how we are to approach the main part of the narrative, the part containing the astronomy of the Moon and the description of the lunar world: we are to read them as allegorical representations, or – to forestall the objection that these allegories are not exactly 'classical' – at least as *imitations* of an allegorical representation.

Since in the case of the lunar astronomy, the commentary presents its underlying signification as the *Dream*'s principal thesis, it is easily identified: The lunocentrism and lunostatism of lunar creatures corresponds to the geocentrism and geostatism of earth-dwellers. The Moon orbiting the Earth thus becomes a figurative representation of the Earth orbiting the Sun.[50]

In the case of the description of the lunar world, matters are more complex. This description, too, should be read as an allegory, but an allegory for which – in analogy to the framing narrative – Kepler does not have a definitive interpretation. In contrast to the frame, however, the reason the true interpretation or translation of this allegory must remain implied is that it is unknown: the right translation would be a description of the lunar world as it is in its actuality, and this is not only beyond Kepler's knowledge, but also beyond his imagination. The world in the Moon cannot be conceived in its potential otherness. It is only imaginable in strict derivation from the terrestrial

[50] What Kepler calls the "thesis of the whole *Dream*," is most explicitly stated in the following passage: "For Levania seems to its inhabitants to remain just as motionless among the moving stars as does our earth to us humans." The comment to this passage explains: "Here is the thesis of the whole *Dream*; that is, an argument in favor of the motion of the earth or rather a refutation of the argument, based on sense perception, against the motion of the earth." (*Dream*, 17 and note 96)

world; as such it can only stand as metaphor. This argument underlies the entire *Dream*.

To make this evident, we must first connect the description of the lunar world, which I have defined as an allegorical representation, with two aspects of the text: the ambiguous correspondences that characterize the structure of the dream narrative and the listing of sources for the narrative material as it is given in the commentary. Here the text reflects on the limitations governing the creative potential of the imagination when it comes to imagining an alternative world by offering a paradigmatic representation of the metaphoric nature of the imagination.

If, therefore, the description of the lunar world is introduced as an allegory, this is done in response to the documentation of the workings of the imagination: as an allegory, the description of the world on the Moon claims to be only allegorically true. It is presented as a metaphorical substitute for an authentic description of this world, which in its potential otherness is beyond the reach of Kepler's imagination. In this way, the text comments on the description of the lunar world as a creation of the imagination and establishes its validity as a representation despite its pictorial inadequacy. The allegorical fable, composed of elements from the material and familiar world, does not represent an abstract concept here, but stands for another *material* world that is beyond knowledge and imagination. And while it may not be totally legitimate to call this allegory, in its structural composition the *Dream* closely *imitates* the structure of allegory, using its imitation of this structure as an initial approach to a mystery that is both beyond knowledge and imagination.

In the summary of the dream narrative, we saw that there is a continuous play of parallelisms and ambiguities between the various levels of narration:[51] the ambiguous

---

[51] Several of the parallelisms referred to here have been pointed out by Timothy Reiss in his analysis of the *Dream*. Cf. "Kepler, His *Dream*, and the Analysis and Pattern of Thought" in *Discourse*, 140-67. This analysis is part of a larger study which attempts "to describe the gradual domination of an earlier '*class*' of discourse by a later one, ... [which] becomes the single *dominant* structure and the necessary form taken by thought, by knowledge, by cultural and social practices of all kinds." (22) Borrowing Lévi-Strauss's concept of the patterning of mythical discourse, Reiss calls the earlier type of discourse one of 'patterning', while he refers to the later type as a discourse of analysis and reference (using the short form of 'analytico-referential discourse'). According to Reiss, Kepler's *Dream* "manifests a moment of transition when [these] two classes of discourse function with equivalent power." (148) They are fairly clearly divided from each other in the text, the discourse of patterning manifesting itself primarily in the prologue and in the conclusion while the analytico-referential discourse occupies the central portion of the text and most of the notes. (144) Reiss argues that the patterns of parallelisms and oppositions apparent in the framing narrative form a discourse of relations which refers not to content but to the ordering of content. (46) It is a discourse that "helps us to the comprehension of a divine order whose nature is ultimately ineffable." (52) Reiss explains Kepler's use of such a discourse with his conviction that "science was a kind of interrogation of the manifestations of the 'world soul' whose essence must forever remain indistinct for humans, by virtue of their own participation in it, but whose limitless appearances were available for ordering into a pattern which would reflect in some way such an essence." (159) This is why the different logic of the dream narrative remains largely intransigent to the analyses performed in the commentary. Accordingly, Reiss' argument regarding the undermining of the allegorical dream as a genre is programmatic for the text as a whole. I would like to show that the pattern of parallelisms, though indeed most conspicuous in the framing narrative, is not limited to this part of the *Dream*, but also manifests itself in the central portion of the text and at

play of identities between the first narrator, Duracotus and the Daemon is coupled with the parallelism between Bohemia's legendary ruler Libussa, Fiolxhilda and the old hags whom the daemons escort to the Moon. The relationship between Duracotus and his mother matches that of the first narrator with Libussa; and when it comes to communicating with the sky, the relationship between mother and son also resembles that between Tycho Brahe and Duracotus. Finally – and most importantly – all of these fictitious characters and their relationships among each other stand in obvious analogy to Kepler's own history.

In addition to the correspondence of characters, we find a correspondence of places: The island Iceland is juxtaposed with the island Hven, where Tycho is engaged in his astronomical observations, and both islands evoke "the island of Levania", which is said to "float" in the sky like an island.[52] The parallelism between Hven and the Moon acquires additional prominence through Duracotus' rough journey to Hven, which all too obviously prefigures the journey to the Moon. At the same time, the Moon is also related to Iceland: for the volcano Mt. Hekla, where Fiolxhilda picks those herbs that are to cause Duracotus' exile, is said to be the entrance to purgatory in Irish mythology.[53] This clearly corresponds to a prominent passage at the end of Plutarch's *De facie*, the myth narrated by Sulla, where the Moon is referred to as the purgatory of souls. Both the allusions to this passage in Kepler's daemonology as well as the fact that Kepler had planned to include his own translation of this text in the printed edition of the *Dream* add significance to this implicit correspondence.

Iceland and the Moon are also linked by the fierceness of their climates and their long days and nights. During these long nights, both the inhabitants of Iceland and the Subvolvans find consolation in the contemplation of their 'moon'.[54] And finally, as the journey to Hven prefigures that to the Moon, the Daemon's description of the Moon is clearly prefigured in Duracotus' relation about his home country, at which Tycho and his assistants are filled with wonder.

I pointed out in the summary that the daemons' flight from the Sun upon their arrival on the Moon into "caves and dark places"[55] parallels the frantic struggle for life of the lunar creatures, who spend half their life underground in an attempt to protect themselves from the scorching Sun. But then Kepler also explains that "the allegory

---

times even extends to the notes. The extension of the pattern to both the description of the Moon and the commentary is, in fact, of central importance to the poetics of the imagination which the text conveys through its structure.

52 *Dream*, 15 and note 54. – "natat potius". (*GW*, 339)

53 *Ibid.*, note 2.

54 On telling her son about her communication with the Daemon, Fiolxhilda explains that in spite of the discomforts caused by cold and darkness, there are also benefits to be reaped from them: "At our service are very wise spirits, who detest the bright light of the other lands and their noisy people. They long for our shadows, and they talk to us intimately. Among them there are nine chief spirits. Of these, one is especially well known to me." In a note at the end of this passage Kepler adds: "I am sure that here I had in mind either Urania among the Muses or astronomy among the sciences. For if the northerners were not deprived of many aids to their life on account of the cold, I should say that they are better suited for astronomy than others, because among them there are greater differences between day and night, and these are phenomena conducive to astronomy." (*Dream*, 14 and note 36)

55 "speluncas et loca tenebrosa". (*GW*, 324)

compares ... the time spent in the caves, to continuous speculation based on observations of the eclipses."[56] He then compares the daemons in the caves to himself in his underground beer cellar, from which he used to conduct his observations. This comment leads back to Duracotus and Fiolxhilda, who are listening to the Daemon's narration while their heads are covered with their clothing. This episode, Kepler explains, was in its own turn inspired by an incident that had happened shortly before he conceived the plan of the *Dream*: While observing a solar eclipse in the emperor's garden, where they lacked a dark room, Kepler and his audience "covered [their] heads with [their] coats and kept out the daylight in that way."[57]

The complex correspondences of characters and settings that echo through the entire narrative of the *Dream* cause the differentiation of both narrative levels and places to collapse: the parallel arrangement of literally all the characters that figure in the narrative[58] (be it with regard to their function in the plot, their character traits or what they experience) combined with the similarities between the various settings (Iceland, Hven,[59] the Moon) seem to imply that the separate parts of the narrative are somehow the same, that they are mere variants of each other. In very simplified terms we could thus say that, just as Duracotus and the Daemon are conceived in the image of the first narrator – and he in turn in that of Kepler, so too is the Moon modeled on Iceland (or rather on the author Kepler's conception of Iceland, based mostly on books). In the imagination of the Icelandic Duracotus (as it is imagined by Kepler), the one serves as basis for the imaginative creation of the other.

Duracotus, his real eyes covered with his clothes, is seeing with his mind's eye instead. One of the notes clearly emphasizes this point when commenting on the appearance of the Daemon, who was "[s]ummoned from the moon, *toward which our eyes were directed in our imagination*."[60] Here Duracotus seems to have taken over the commentary since this comment is his and not Kepler's – unless, of course, it applies to both, which is just what is suggested by the continuous parallelism mentioned two paragraphs earlier. Since characters from all the narrative levels are referred to as sitting in darkness, the opposition between their not seeing or hiding from daylight (Duracotus, the daemons, the lunar inhabitants) and visual observations conducted in

---

56 *Dream*, note 85. – "Sed Allegoria ... comparat ... morae in cavernis, speculationem continuatam, observationibus Eclipsium insistentem." (*GW*, 342, note 82) In two earlier notes Kepler explained that both lunar and solar eclipses offer opportunities to conduct shadow measurements and thereby obtain knowledge of celestial phenomena. (Notes 55 and 56)

57 *Dream*, 14 and note 49. Adaptations mine. – "quare pallijs obnubentes capita, ita diei lucem arcuimus." (*GW*, 338)

58 This even includes the skipper, whose sole function is that of forming the transition from Iceland to Hven: as the conductor of the tumultuous sea journey, he is the counterpart to the daemons, who are responsible for the later transition from Iceland to the Moon.

59 Although Hven also participates in the parallelism, it does not do so with respect to its characteristics as a place – the exception lies in its being an island. Other than this, there are no references to its geographic characteristics. Its primary function is that of constituting a moment of transition: from childhood to adulthood, from ignorance to experience and, most importantly perhaps, from Iceland to the Moon. Thus, it makes sense that the journey to Hven and that to the Moon – both moments of transition in a physical sense – should be presented as variants of each other.

60 *Dream*, note 52. Emphasis and adaptation mine. – "Ex Luna accersita, *in quam oculi per fictionem erant translati*."

the dark (Kepler as astronomer, the allegorical interpretation) is turned into an analogy. As the analogy suggests, the description of the world in the Moon in the second part of the *Dream* is not based on any visual evidence, but rather results from groping in the dark: As a fiction evoked from Duracotus' imagination, dreamt by the first narrator, and ultimately created in the imagination of Kepler, the author, it is thoroughly solipsistic in as much as it is made up entirely of the recycled contents of Kepler's mind – his 'storehouse of images'. This idea resounds through the hollow structure of the text. It is also reflected in the passage where the Daemon describes how the shapes of the continents on the terrestrial globe appear to the observer on the Moon:

> In the darker half the shape of the spot is hard to describe. Yet on the eastern side [toward the Atlantic Ocean] it looks like the front of the human head cut off at the shoulders [Africa] and leaning forward to kiss a young girl [Europe] in a long dress [Sarmatia, Thrace, the Black Sea regions, Muscovy, Tartary], who stretches her hand back [Britain] to attract a leaping cat [Scandinavia]. The bigger and broader part of the spot [Asia], however, extends westward without any apparent configuration. In the other half of Volva the brightness is more widely diffused [the two oceans] than the spot [the American continent]. You might call it the outline of a bell [South America] hanging from a rope [Nicaragua, Yucatan, Popayan {i.e., Colombia}] and swinging westward. What lies above [Brazil] and below [North America] cannot be likened to anything.[61]

> Figura difficilis est explicatu. Parte tamen orientaliori cernitur instar Protomes capitis humani, axillarum tenùs resecti, admoventis ad oscula puellulam cum veste longa, quae extenta retrorsum manu felem assultantem provocet. Major tamen et latior maculae pars sine evidenti forma versus occidentem procurrit.
>
> In altera medietate Volvae, latiùs diffunditur splendor quàm macula. Effigiem diceres Campanae de fune dependentis et in occidentem jactatae. Quae supra infraque sunt, assimilari nequeunt.[62]

The notion 'Europe gives Africa a kiss' was current at the time. There are quite a number of fantastical illustrations of Europe from the sixteenth century where Europe is represented as a royal princess with Spain figuring as her head (see *fig. 1*). One such image was offered to Charles V as a gift.[63] Besides the fact that Kepler is making use of a courtly emblem charged with political significance to describe a viewpoint that purports to be extraterrestrial and thus detached, what is noticeable about Kepler's adoption of this image in the Daemon's narrative is the coincidence of the girl's head with Spain. This corresponds in a peculiar way to another comment in the notes where Kepler elaborates on the Daemon's statement that "the firm bodies of Spaniards" are particularly suitable for the strenuous flight to the Moon. Kepler himself praises the

---

61 *Dream*, 24. The explanations in square brackets are Kepler's own given in the notes. Note that east and west, upper and lower part are reversed in the lunar perspective. Moreover, at the time Kepler wrote the dream narrative he mistakenly believed water to have a higher albedo than land, which is why he describes the oceans as lighter areas and the continents as 'spots'. Kepler acknowledges this mistake in note 154 as well as in his *Conversation* and credits Galileo for enlightening him on this matter.

62 *GW*, 328.

63 Cf. *Keplers Traum vom Mond*, ed. and trans. Ludwig Günther (Leipzig, 1898), editor's comment on Kepler's note, 114, note 115.

AFRICA
OCEANVS
MARE MEDITER.
ANGLIA
GALLIA
GERMANIA
DANIA
VANDALIA
SICILIA
VNGARIA
POLONIA
LITHVANIA
MACEDO
BVLGARIA
GRAECIA
SCYTHIA
ASIA

figure 1

Spaniards' great intellects, their fine judgment as well as their asceticism, which makes for great astronomers.[64] Together with the Daemon, he deplores the gluttony of the Germans, which renders them completely unsuitable for an exact science like astronomy. Accordingly, Germany – or the holy Roman Empire – in the Daemon's description coincides with the girl's stomach.[65]

At the same time, it is remarkable that all the territories on the terrestrial globe which the Europeans were only just beginning to explore and map have no meaningful shape for Kepler. Similarly, of all the regions he characterizes all except Europe are described as inanimate objects (Asia as the girl's dress, Africa as a chopped-off head, South America as a great bell).[66] The Daemon's emblematic characterization of the spots on the terrestrial surface thus not only represents a distinctly European viewpoint, it also humorously reflects Kepler's personal prejudices. And although in its Europeanness the Daemon's description may cater well to a European audience, its Keplerian bend lends the whole passage a decidedly ironic quality.

One character is conceived in the image of another character, this world furnishes the source material for the representation of that world – what the *Dream* implies in schematic terms through parallelisms both of structure and content, it offers in emblematic form in the motif of the dream. I mentioned earlier that the *Dream* contains a paradigmatic representation of the formation of a dream within its narrative; the narrative marks (implicitly at least) the psychophysiological factors responsible for the dream's form by referring to the substance from which the dream experience takes its expression. References of this kind occur, first of all, at the very beginning of the nar-

---

64 The Daemon explains: "No men from Germany are acceptable; we do not spurn the firm bodies of Spaniards." In the note added to this passage Kepler comments: "In thinking up jokes, you should bear this, too, in mind, that while you believe you are winning the approval of one listener within the hearing of another, you deeply offend the latter and also his neighbor. Even so, Germany is just as famous for corpulence and gluttony as Spain is for genius, discernment, and temperance. Therefore in the exact sciences, the category to which astronomy belongs (especially this lunar astronomy, which stands in an unusual position if an observer looks out from the moon), should a German and a Spaniard exert themselves equally, the latter would come out far ahead. And that is why I predicted that this book would be laughed at by the Germans, but held in some esteem by the Spaniards." *Dream*, 15 and note 61.

65 Indeed, in the print reproduced here, Bohemia figures as the princess's stomach!

66 The representation of Africa, moreover, clearly reveals how little was known about this vast continent: Whereas the coast is described in minute detail in the image of a face, the rest of the continent's shape is blurred. This reflects the perception of Africa at the time as it is represented on contemporary geographical maps, where it figures mostly as an indistinct landmass which counterbalances the landmass of Europe and Asia, or is given a fantastical shape. Cf. Leo Bagrow, *History of Cartography*, 2nd ed. revised by R.A. Slelton (Chicago: Precedent Publishing Inc., 1985). Philip Stooke points out cases of early maps in which the shape of Africa shows no resemblance to its real form, but bears similarities to one of the Moon's 'ancient' spots, so as to suggest that cartographers might have based their drawing of maps on the theory that the spots on the Moon reflect the shapes of the continents on Earth. They may accordingly have used the Moon's spots to guide them in the depiction of territories on Earth that were still largely unexplored. Cf. "Mappaemundi and the Mirror in the Moon," in *Cartographia* 29, no. 2 (1992): 20-30. In the *Conversation*, Kepler relates that his patron, the Emperor Rudolph II, still believed in this theory and, on looking at the Moon through a telescope, was convinced that he could make out the reflection of Italy on the lunar surface. (13)

rative, where the narrator describes the situational context from which the dream originates. As Fernand Hallyn puts it:

> Kepler's *Dream* is presented as the product of the occupations of the day, since it melds into a single narrative the 'residue' of the reading of a magician's story and observations through a telescope. Elements originating from a distant past are also grafted on this outline: the lack of a father, the visit with Tycho Brahe, memories of Prague ....[67]

Just as the prologue offers hints about the dream's formation by presenting the material from which it takes its shape, so does the abrupt end: Here the dreamer's immediate physical situation reveals obvious correspondences to episodes in the dream so as to suggest that this is where these episodes originated from in the first place. Hallyn has observed that "Aristotle had already drawn attention to the intensification and the transformation that the dreamer's imagination is subject to when presented with nervous stimuli."[68] Let us see how this works in the *Dream*: from the image of rain prevailing over the greater part of the Subvolvan hemisphere, the first narrator is roused from his dream by a wind and "the rattle of rain" only to find that, in analogy to Duracotus and Fiolxhilda, his "head [was] really covered with the pillow and [his] body with the blankets."[69] Accordingly, wind and rain, pillow and blanket act as stimuli for analogous situations in the dream narrative.[70]

On the level of the fiction, the narrative thus documents the genesis of the dream experience it relates. However, since the dream motif figures as an emblem, the docu-

---

67 *Poetic Structure,* 265. The word *residue* is placed in inverted commas because it is quoted from a passage in Cicero concerned with the causation of dreams. Hallyn argues that most theorists concerned with the phenomenon of dreams took into account psychophysiological factors that could determine a dream's form by using a typology borrowed from Aristotle, Hippocrates, and Galen. He shows that what Kepler offers in the *Dream* is, among other things, an illustration of a long tradition of theoretical inquiries into the nature of dreams.

At the beginning of the *Somnium Scipionis*, the protagonist voices his belief that the dream he is about to relate was triggered by a conversation he had shortly before he went to sleep: "The following dream came upon me, prompted, I suppose, by the subject of our conversation; for it often happens that our thoughts and words have some such effect in our sleep as Ennius describes with reference to Homer, about whom, of course, he frequently used to talk and think in his waking hours," in *De re publica*, trans. Clinton Walker Keyes (Cambridge, MA: Harvard University Press, 1966), vol. 6, x, 263.

68 *Poetic Structure,* 267.

69 *Dream*, 28-29. Adaptation mine. For the original text, see note 23, *supra*. The text makes the correspondence between the narrator's physical situation and that of Fiolxhilda and her son very clear in that it refers once again to the fact that their heads are covered up: "Therefore, leaving behind the Daemon narrator and her auditors, Duracotus the son with his mother, Fiolxhilda, as they were with their heads covered up, I returned to myself and found my head really covered with the pillow and my body with the blankets. It may be worth pointing out that, in his translation of this passage from the Latin, Edward Rosen refers to the Daemon as female, thus setting 'her' parallel to Fiolxhilda rather than Duracotus. The Latin version leaves the question of the Daemon's gender open and thus offers a further source of ambiguity. The Greek word 'daimon', after all, can take both the masculine and the feminine gender.

70 Hallyn mentions that the dreamer's pillow and blanket act as nervous stimuli that shape the dream experience. The point about wind and rain has been added here.

mentation of the dream's genesis must have a broader application. We established earlier that the activity of dreaming is analogous to that of imagining and that, in simple terms, dreaming could be defined as a particularly vivid, unconsciously performed act of the imagination.[71] If Kepler's narrative offers a documentation of how the dream about the Moon is generated, it is likely to offer a similar documentation of how the dream narrative is created in the author's imagination – and so it does: In analogy to the narrator's dream experience, where indications about its formation are given in the margins of the dream narrative, the documentation of the *Dream*'s production as a creative process of the imagination appears in the margins of the text, in the commentary.

We have already noted that the commentary emphasizes the textuality of the *Dream*, thus characterizing it as a literary creation. As such, it also characterizes it as a creation of the imagination.[72] While the fictitious dream in the narrative is shown to develop from the dreamer's more or less immediate context, the literary *Dream* presents itself as the product of a mental process in which both memories of the author's personal experiences as well as knowledge and information he acquired through extensive reading or hearsay are adapted and blended together. The process of this creative act is painstakingly documented in the notes that follow the text. The commentary leaves no doubt about the fact that every single element in the narrative is a quotation from the mental storehouse of Kepler's past experiences. The commentary thus documents what is schematically presented through the correspondences between the narrative levels and rendered in emblematic terms through the dream motif.

This documentation is especially rich in reference to the framing narrative and the description of the lunar world; in the case of the lunar astronomy, it is confined to references to astronomical observations and to explanations of the mathematical and geometrical calculations underlying the celestial phenomena the Daemon describes. The reason for this is obvious: whereas an astronomy of the Moon essentially relies on the visual evidence available and is produced by applying visual evidence and optic measurements to a cosmological framework, for a hypothesis regarding the specific qualities of the lunar world such astronomical data can only offer a basic guideline; from there, it is up to the imagination to create a world in graphic detail. For this reason, an inquiry into the process of imaginative creation and its foundation is particularly relevant in the case of the description of the lunar world and the forms of life it accommodates. That is to say, such an inquiry is of special interest when there are no direct human experiences on which to base this description.

Once again, as with the allegory, the particular focus given to the framing narrative and the process of its composition sets the focus for the description of the lunar world, increasing our awareness of the provenance of its constituting images: On the principle that both on the Moon and on Earth forms of life are adapted to circumstances, Kepler creates a lunar habitat that directly derives from phenomena on Earth which he has read about and which he deems comparable in terms of climatic conditioning. It is a world that consists entirely of quotations. This fact is carefully documented in the notes, which refer to the analogous terrestrial phenomena and indicate the textual sources that describe them. In the following, I will highlight some of these correspon-

---

[71] For this, see p. 77, *supra*.
[72] *Ibid.*

dences by offering a reading of the description of lunar life with reference to Kepler's comments.

Within the dream narrative, the description of a lunar habitat is mostly limited to the Privolvan hemisphere, where conditions for life are harsher mainly owing to the absence of the Volva. As Kepler emphasizes in a note, this is the part of the lunar globe "[w]hich no eye has ever seen." Yet he promises to defy the limits of visibility with his "sound reasoning" (*rationem*).[73] The atrocity of the Privolvan climate makes a settled form of existence impossible to the effect that "in general the Subvolvan hemisphere is comparable to our cantons, towns, and gardens; the Privolvan, to our open country, forests, and deserts."[74] "In the course of one of their days [the Privolvans] roam in crowds over their whole sphere, each according to his own nature."[75] They spend their lives in constant flight from the Sun, speeding with the receding tide to find relief in the coolness of the deep waters or seeking protection underground, where the heat of the Sun cannot reach. Kepler corroborates this point with a comparison to "our cellars and deep wells, in which we keep our drinks cool in the summertime."[76] It may be worth pointing out that this analogy serves as evidence for Kepler's thesis that (sun)light is associated with heat;[77] this association leads Kepler to conclude that the lunar surface, which at certain periods is closer to the Sun than the terrestrial surface and is exposed to its rays for far longer stretches of time, undergoes extreme changes of temperature.

The physiology of Privolvan creatures is adapted to these conditions: Some have legs "which far surpass those of our camels",[78] some resort to wings, while others, helping nature with art, follow the receding water in boats. "Most of them are divers; all of them, since they live naturally, draw their breath very slowly; hence under water they stay down on the bottom."[79] Kepler again confirms the viability of this postulate in the commentary, this time with a reference to "Cola, the Sicilian man-fish."[80] Edward Rosen, tracing this reference to Julius Caesar Scaliger and from there to Jerome Cardan, explains that this name was given to a young diver who, like a fish, used to hide under water for three or four hours. Cola, of course, is part of an old legend, a fact

---

73 "Quos oculus nullus unquam vidit." (note 210)

74 *Dream*, 28. – "Nam in universum Subvolvanum Hemisphaerium nostris pagis et oppidis et hortis aequiparatur: Privolvanum nostris agris et sylvis et desertis." (*GW*, 330) The distinction between a civilized and an uncivilized world also reflects the contemporary distinction between Europe and Asia, on the one hand, and the new territories, on the other hand. In harmony with this distinction, many of the creatures that populate the Privolva are based on reports about remote territories and the unfamiliar forms of life to be found there.

75 *Dream*, 27. Adaptation mine. – "Totum globum unius suae diei spacio agminatim pervagantur ... prout cuique natura est." (*GW*, 330)

76 *Dream*, 28. – "eum ego ad nostras cellas, et puteos profundos remitto, in quibus aestate, potum nostrum infrigidamus." (note 217)

77 For this, see Rosen's note 357.

78 "quibus longissimè camelos nostros praevertunt". (*GW*, 330)

79 *Dream*, 27-28. – "Urinatores sunt plurimi: omnia naturaliter animantia spiritum trahunt lentissimè: sub aquis igitur degunt in profundo". (*GW*, 330)

80 "Cola Siculo, Homine-pisce". (note 214)

which, according to Rosen, Kepler's reverence for both his sources prevented him from realizing.[81]

The narrative continues: "In plants, the rind; in animals, the skin, or whatever replaces it, takes up the major portion of their bodily mass; it is spongy and porous. If anything is exposed during the day, it becomes hard on top and scorched; when evening comes, its husk drops off."[82] Once again, the commentary explains: "The precedents were our vegetables' and fruits' rinds, varying with nature's varying providence; the precedents were oysters' and turtles' shells, shaped like shields; the precedents were the calluses on the feet, the hoofs and soles of animals."[83] By referring to these physical characteristics of animals and plants, Kepler first of all indicates the inspirational source that fed his imagination in the creation of lunar life. Yet he also draws attention to the fact that the physical characteristics of animals and plants on Earth ought to be understood in the context in which each animal lives and each plant grows: Every living thing on Earth has been created in adaptation to its environment. Therefore, "nothing on earth is so fierce that God did not instil resistance to it in a particular species of animals: in lions, to hunger and the African heat; in camels, to thirst and the vast deserts of Palmyrene Syria; in bears, to the cold of the far north, etc."[84]

The adaptation of different forms of life to their environment constitutes a central principle in God's creation. It also turns out to be the main principle in Kepler's imaginative creation of Privolvan forms of life. In fact, it is remarkable how Kepler's creative act is set parallel to its original – in imitation, so to speak. Accordingly, in the passage immediately preceding the lines just quoted, Kepler argues:

> Since I had deprived the Privolvans of water, and I was compelled to leave them immense alternations of heat and cold coming directly on each other's heels, it occurred to me that those regions could not be inhabited, at least not out in the open. Hence it was convenient that water flowed in at fixed times of the day [Kepler is referring to the tides]. When it receded, I had the living creatures accompany it. To enable them to do so quickly, I gave some of them long legs and others the ability to swim and endure the water, with the proviso that they would not degenerate into fishes. None of this will be unbelievable to anybody who has read about Cola, the Sicilian man-fish.[85]

---

[81] Cf. Rosen's note 353.

[82] *Dream*, 28. – "In stirpibus cortices, in animalibus cutis, aut si quid ejus vicem obtinet, majorem corporeae molis partem absolvit, fungosumque et porosum est; ac si quid deprehensum in die fuerit, in summitate induratur et aduritur; vesperâ succedente deglubitur." (*GW*, 330)

[83] "Praeivêre nostrarum stirpium et fructuum cortices diversi, diversâ provisione naturae, praeivêre testae Ostrearum et testudinum clypeiformes: praeivêre calli pedum, ungulae et soleae animantum." (note 218)

[84] "nihil in terris tàm esse nobis violentum, cujus tolerantiam non indiderit Deus certo generi animalium; famis et aestus Africani, Leonibus; situs et immensorum desertorum Syriae Palmyrenes, Camelis; frigoris Hyperborei, Ursis, etc." (note 214)

[85] Explanation mine. The remark about the 'proviso that they would not degenerate into fishes', it seems to me, is a humorous reference to a passage in Plato's *Timaeus* where the evolution of women, birds, land animals and fish is explained in terms of a process of degeneration from the male human being. "And the fourth kind, which lives in the water, came from the most utterly thoughtless and stupid men, whom those that remoulded them deemed no longer worthy even of pure respiration, seeing that they were unclean of soul through utter wickedness; wherefore in

> Cùm aquas Privolvis ademissem, aestûs et frigoris vicissitudines immanes proximis omninò temporum articulis relinquere cogerer: subijt animum, eos tractus habitari non posse, saltem sub dio. Ex opportuno igitur aquae influebant certis diei tempestatibus: eas iterum recedentes jussi viventia comitari; et ut tam properè possent, longos pedes dedi; alijs nandi facultatem, et aquarum patientiam: ut tantùm non in pisces degenerarent. Nec quicquam hujus illi erit incredibile, qui de Cola Siculo, Homine-pisce, legerit.

The two creations must, of course, be considered in due proportion, a point we will return to later on in relation to the allegorical quality of Kepler's description. Nevertheless, when Kepler tries to imagine ways in which life on the Moon might be conceivable, he does this in clear imitation of those principles in Creation that are manifest to the human contemplator.

Kepler offers alternatives: Besides making the Privolvans physically fit to find protection from the heat of the Sun, he proposes the possibility of their full adaptation to the climatic conditions on the Moon. Proceeding with the disjointed logic of dreams, the narrative thus suddenly declares: "In general, the serpentine nature is predominant. For in wonderful manner they expose themselves to the sun at noon as if for pleasure".[86] The Privolvans have here turned into sun-worshippers. And together with all the other characteristics with which they are endowed, their affinity to the Sun draws on a terrestrial precedent; Kepler relates it to a passage he (mistakenly) remembers having read in Arnobius the African:

> that among the pleasures of his people it was an accepted practice to expose themselves naked to the sun, basking like lizards, and, *unless I am mistaken*, also like crocodiles. The nature of the latter as beasts of the country is taken on by the Africans, I supposed, on the basis of this similar practice. For it would surely be rather a sort of torture for us Europeans.[87]

> [Legeram in Arnobio Afro,] inter delicias ejus gentis et hoc receptum, ut se nudos exponant Soli, apricantes ut lacertae, et, ni fallor, etiam crocodili; quorum, ut bestiae patriae, naturam Afros habere ex eo censui. Nam id quidem nobis Europaeis tormenti potiùs genus esset.[88]

The difference in physiology between 'us Europeans' and 'the Africans', which is attributable to the difference in climate, is here made analogous to that between 'us

place of air, for refined and pure respiring, they thrust them into water, there to respire its turbid depths. Thence have come into being the tribe of fishes and of shell-fish and all creatures of the waters, which have for their portion the extremest of all abodes in requital for the extremity of their witlessness." (253)

86 *Dream*, 28. – "Natura viperina in universum praevalet. Miraculo enim sunt, Soli sese in meridie exponentes, quasi voluptatis causa". (*GW*, 330)

87 Note 219. Emphasis mine. Rosen points out that Kepler is misquoting from memory, for Arnobius did not say that nude sun-bathing was an accepted practice of his people. Rosen adds that "[i]n a private letter dated January 17, 1605, Kepler remarked with equally mistaken assurance that according to Arnobius, Africans are akin to reptiles in nature." Rosen's notes 356 and 257. Adaptation mine.

88 Adaptation mine.

Europeans' and the inhabitants of the Privolva, thus serving not only to make the possibility of life on the Moon imaginable in defined pictures, but to make it conceivable in principle. Besides the fact that Kepler is remembering wrongly, what is interesting about this reference is that he should be drawing attention to the circumstance that he is quoting from memory and that his memory may be playing tricks on him.

The African fable is followed by one about the people of "Lucumoria" in the far north, which Kepler attributes to the Jesuit theologian Martin del Rio and which relates "that some of [these people] die when they are assailed by that long night, and revive when the sun returns".[89] In the case of the Privolvans, Kepler reverses this process:

> To certain of them the breath they exhaust and the life they lose on account of the heat of the day return at night; the pattern is the opposite of that governing flies among us. Scattered everywhere on the ground are objects having the shape of pine cones. Their shells are roasted during the day. In the evening when, so to say, they disclose their secrets, they beget living creatures.[90]
>
> Quibusdam per diei aestum spiritus exhaustus, vitaque extincta, per noctem redeunt, contrariâ ratione quàm apud nos Muscis. Passim per solum dispersae moles figurâ nucum pinearum, per diem adustis corticibus, vesperi quasi reclusis latebris, animantia edunt.[91]

The fabulous account of the Lucumorians, which forms the background for the first part of this passage, is blended in the second part with yet another legend that was first reported by Scaliger and is retold in the commentary:

> From the resin which exudes from ship timbers as a result of the sun's heat and sticks together in a ball, ducks are born. The last part of their whole body to develop is the bill. When this is released, they slip into the water below. So says Scaliger in his *Exercises*. The Scottish tree which produces the same offspring is well known because many people have made it famous.
>
> Ex resinâ, exsudante ex trabibus navium per Solis fervorem, et globatim adhaerescente, Anates nasci, quibus ultimum totius corporis, rostrum maturescat, quo soluto, se dent undis subjectis, refert Scaliger in Exercitationibus. Nota est multorum celebratione, arbor Scotiae, quae eundem foetum proferat.[92]

In his *Nautical Diary*, Gerrit de Veer exposes this biological conception theory as a legend and offers an explanation for its formation: the ducks Kepler is referring to are called "barnacles" or "*Rotgansen*". Since these birds lay their eggs and raise their young in Greenland at a latitude of 80°, their reproduction had been a mystery up to that point.

---

89 "esse qui, nocte illâ longâ ingruente, emoriantur, Sole redeunte, reviscant". (note 220)

90 *Dream*, 28. Adaptation mine. The reference to the return of Kepler's flies to life is as fabulous as the other stories. Rosen suggests that Kepler is here "somehow further muddl[ing an] already muddled account in Pliny's *Natural History*." Cf. Rosen's note 359. Adaptation mine.

91 *GW*, 330.

92 *Dream* and *GW*, note 221.

> This is the reason why some authors were not afraid to write that in Scotland the barnacles are born on trees whose branches grow out over water; if the fruit drops into the water, goslings are born which swim at once; but if the fruit falls on the ground, it rots and does not develop. The falsity of these statements is now clear.[93]

As Rosen points out, Kepler had studied this brief book carefully and could therefore scarcely have overlooked the passage. "Nevertheless, he chose to repeat the medieval legend of the barnacle goose, and to ignore ... [de Veer's] historic refutation".[94] Once again, Rosen bases his argument on the assumption that the principal reason why Kepler used this story in his narrative is that he believed it to be true.

In "Appendix M" of his edition of the *Dream*, Rosen offers an interesting documentation of a similar if more complex situation concerning Kepler's account of the population of Lucumoria. He notes first of all that Kepler altered del Rio's account and that he misspelled the name of the Russian district, which in del Rio appears as "Lucomoria." Rosen calls this "an excellent example of Kepler's frequent practice of embroidering his sources, just as the garbling of the name shows that like most of his contemporaries he did not always take the trouble to verify his quotations."[95] Kepler is thus again quoting from memory and taking certain liberties with his material. In addition, Rosen observes that del Rio's is only the third link in the chain of accounts of this resurrection story: The Jesuit openly acknowledged having adopted it from a certain Alessandro Guagnini and added that the same story had also been related by "Sigismund the free baron", who in turn had been Guagnini's source. This baron Rosen identifies as Sigmund von Herberstein (1486-1566), the Holy Roman Empire's ambassador to the Czar of Russia and author of *Commentaries on Muscovite Affairs*, where the resurrection story is related. Rosen points out that Kepler was not only personally familiar with this book since he had relied on it elsewhere, but that he knew the Herberstein family very well.[96] Herberstein himself had characterized the resurrection story as "just like a fable", a judgment that del Rio reported in his own version of the story. Kepler, therefore, must have been well aware of the disclaimer.

Rosen argues that even though Kepler, presumably, had noticed the passage in which Heberstein expressed his disbelief regarding the truth of this story, he paid no attention to it because the tale "unquestionably appealed to one side of his personality." To Kepler the astronomer, who seeks the physical causes of phenomena, Rosen opposes Kepler the theologian, who is "favorable to doctrines for which no cause can be assigned."[97] Just as with the legends about the barnacle geese and the man-fish Cola,

---

[93] Quoted in Rosen's note 362.

[94] *Ibid.* Adaptation mine.

[95] *Dream*, 236.

[96] "The Herbersteins were the leading noble family in Graz, where Kepler spent six eventful years. He dedicated his *Cosmographic Mystery*, which he wrote while he was living in Graz, principally to 'Sigmund Frederick, the free baron of Herberstein'. This Sigmund Frederick, Freiherr von Herberstein (d. 1621), was not the son of the Sigmund Herberstein who wrote the *Commentaries on Muscovite Affairs*. ... The two Sigmunds belonged to different branches of the same Herberstein family, which enjoyed great renown throughout Austria." (*Ibid.*, 237)

[97] The quotation is taken from an observation Kepler himself had made about his friend Fabricius, a theologian. Rosen argues that "[d]espite [his] immense contribution to the advancement of science,

Rosen thus tacitly assumes that the only way to explain Kepler's use of such fabulous materials is to say, first of all, that he believed them to be true and, secondly, that he expected his audience to believe the same.

While Rosen's argument may be valid as a general characterization of Kepler's mentality, which in a peculiar way combined a scientific mind with a predilection for the mythical, the magical and the transcendental, as an explanation for Kepler's choice of illustrative materials in the *Dream* it is hardly satisfactory. First of all, it does not account for the following questions: Why, since he was familiar with Herberstein's account, did Kepler attribute the story about the Lucumorians to del Rio even though the latter acknowledged having copied this story from two other sources? And why, if his aim was to convince, did he refer the reader to stories which he knew had been exposed as myths – and this not only once, but twice in immediate succession? Moreover, Rosen's explanation does not consider Kepler's programmatic use of fabulous accounts in the wider context of the *Dream*, where the narrative presentation repeatedly stands in a contradictory relationship to the scientific thesis proposed.[98]

Whether or not Kepler truly believed in Herberstein's account or any of the other fabulous accounts recycled in his description of lunar life must, I think, remain an open question. And even while the question as such is certainly legitimate, in the specific context of the *Dream* it is negligible, for here the focus is on techniques of representation. The various stories Kepler recycles in his narrative account of Privolvan life (the resurrection of the Lucumorians, the biological conception of barnacle geese, the Africans' reptilian physiology, Cola's adaptation to life under water) are not used for their respective truth value, but for their ability to illustrate a basic principle: the principle that forms of life are adapted to their environment. This principle, Kepler's scientific argument runs, underlies the entire Creation and makes its individual features fundamentally comparable.

When Kepler attributes the story of the Lucumorians to del Rio rather than to Herberstein, he not only refers to the story itself, but to its narrative recycling, its re-

---

Kepler still retained traces of the mental attitude which he noticed in his theologically-minded friend, Fabricius." (*Ibid.*, 238, adaptations mine)

98 Fernand Hallyn has made a point of the fundamentally oxymoronic nature of the *Dream*, in which hermeticism, magic and demonology constitute the frames for Kepler's defense of Copernican theory. As an example, he refers to Kepler's use of the Earth's *umbra* that serves the daemons as a ladder for their journeys back and forth between the Earth and the Moon. Hallyn observes that this episode relies on the Pythagorean myth which identified the Earth's *umbra* with Styx while the Moon was the abode of the souls of the deceased. (The myth is recounted in the final part of Plutarch's *De facie*.) Pythagorean demonology thus clearly presupposed a geocentric universe in which the Moon is located on the frontier between the physical and the metaphysical world and associated with beings that are intermediary between gods and men. "The *Dream* contains this paradox: while one of its proclaimed goals is to promote the new concept of a heliocentric universe, the fiction that overlays this defense borrows some of its materials from the ancient geocentric cosmology. ... the geocentric universe, scientifically unacceptable as it had become, nonetheless remained more satisfactory in terms of imaginative associations, identifications, and correspondences." *Poetic Structure,* 264-65 and 279-80. Despite the acuteness of this observation, however, Hallyn attributes the peculiar choice of narrative materials in the *Dream* only to one of Kepler's typical whims and argues that like Kepler the "*Janus bifrons* ..., the *Dream* looks both to the past and future in a sustained oxymoron." Hallyn's explanation is thus in the same vein as the one Rosen offered for the reference to Lucumoria.

presentation. The story's own history thus effectively mirrors the generative process of Kepler's lunar description by repeating this process indefinitely. This is true in that Herberstein's account is by no means the beginning of this history: the baron, too, adopted the story from another source, this time a Russian travel book.[99] Besides mirroring Kepler's method of composition with regard to the description of the lunar world, the history of the story's retelling is also a *mise en abîme* of Kepler's larger narrative and the way it is organized into several interlocking narrations. Both in the case of the dream narrative as well as in that of the story about the Lucumorians, the original source of the account not only is constantly deferred, but remains vague: In one case it has its beginning in sleep, in the other in death.

The recycling of a story implies a certain distance between the narrator and the story's content: The story is taken from somewhere else, is only borrowed. This implicit distancing is emphasized by the fact that Kepler at least twice refers his readers to accounts which he knew had been exposed as myths. The tales that become Kepler's models are merely used as narrative material; as such they need not necessarily be true – they need not even be believable. They simply serve to represent something else. It is in this light that we should read Kepler's acknowledgment that in referring to his sources he is quoting from memory and that his memory may be faulty. The degree of correspondence between his quotations and the sources they originate from is thus left open.

However, because of its firm rooting in terrestrial myths and imagery or in a completely terrestrial reality, Kepler's description of life on the Privolva is denied the right to represent truly, that is, literally. Accordingly, his lunar world lays no claim to being a realistic description of the world on the Moon, it only takes its place: Through the focus provided by the dream narrative, Kepler's description is offered as an allegorical representation of a world that remains unimaginable in its otherness.

Yet why should Kepler offer his description of the lunar world in the form of an allegory and how does his narrative imitate an allegorical discourse? We noted earlier on that the failed allegorical interpretation of the framing narrative calls attention to the most basic characteristics of allegory as a trope: the substitution of one discourse for another, the composition of the allegorical narrative and the translation from the literal discourse into the one implied. These points can now be applied to Kepler's description of the lunar world: (1) allegory typically involves the substitution of a graphic discourse for an abstract or transcendental level of signification. Accordingly, Kepler's lunar fiction with its firm rooting in his personal experiences is substituted for the unfamiliar world in the Moon. (2) The odd mixture of fiction and fact on which his narrative builds is shown to be composed of images from the mental storehouse of Kepler's past experiences, that is, from images accessible to his imagination. That this is its distinctive characteristic is abundantly documented in the commentary. (3) Allegory,

---

99 "Whatever I have reported up to now has been literally translated by me from a Russian travel book, which was offered to me. Among these reports, some seem to be legendary and scarcely credible, for instance, the one about men ... who die and return to life. ... Although I myself have carefully investigated these reports, I have been unable to learn anything definite from anybody who has seen these things with his own eyes (although everybody says they were true). Nevertheless, in order to afford others an ampler opportunity to investigate these matters, I decided to omit nothing." (Quoted by Rosen in *Dream*, 238)

moreover, implies the translation from the literal level of signification to the one intended. Accordingly, the allegorical nature of Kepler's description of a lunar world implies the translation his description would have to undergo in order to become truly representational.[100]

For all its inaccuracy, Kepler's lunar world thus nonetheless purports to contain the true lunar world by implication, as a dormant potential so to speak. And here lies the final goal of Kepler's allegorical program: in adumbrating what the literal narrative cannot name. As an allegory, Kepler's lunar narrative points beyond the limited inventory of images accessible to his imagination. However, in so far as the translation from the literal sense into the one intended must remain implied because its true identity is unknown, the description of the world in the Moon is not a genuine allegory, but is offered as an imitation of an allegorical representation.

If the literal discourse is to be read as an allegory, there needs to be an evident relation between the literal sense and the one implied where the potential for translation that the allegory entails can manifest itself. This relation is established, first of all, with the claim that the lunar world, including the kind of life it may accommodate, reflects the Moon's cosmological position and the climatic conditions this position is likely to produce. In other words, the allegory is related to the true image through two basic principles: first, the principle of environmental adaptation, which, Kepler presumes, underlies any form of life both on Earth and beyond, and, secondly, the corresponding and more fundamental principle which holds that just as "nothing on earth is so fierce that God did not instil resistance to it in a particular species of animals", so any part of the universe that is comparable to the Earth in nature is designed to accommodate life.[101] The potential of Kepler's allegory to point beyond its appearance thus depends on the premise that the terrestrial and the lunar worlds are fundamentally comparable with each other.

This need for comparability is also evident in a passage concerned with the near side of the lunar globe. The passage appears in a different part of the text, separated from the dream narrative; for just as the description turns to the Subvolvan hemisphere, the first narrator awakes, thus bringing the dream narrative to an abrupt end. Kepler's thesis regarding forms of life on the Subvolvan hemisphere is presented in an appendix – as an afterthought. And an afterthought it is, for Kepler's theory of circular fortresses, built by the rational Subvolvans, is based on observations he conducted with the telescope in 1623, many years after he wrote the narrative. At the same time, it is by not including this thesis in the dream narrative that Kepler is making a stronger claim for the correctness of his theory, which, contrary to the description of the Privolvan hemisphere, is based on visual evidence: the circular spots on the lunar surface first discov-

---

[100] This translation is comparable to the one Kepler undertakes when he explains the factual basis of the myth underlying Fiolxhilda's magical wind bags. Cf. pp. 79-80, *supra*. Moreover, I pointed out earlier that – if we follow the fiction – the Daemon's account of a lunar astronomy and world as we read it in the *Dream* involves at least one translation: since it was originally given in the Icelandic language, the Latin version in the *Dream* must be either a translation by Kepler or Duracotus. The translation into Latin makes the Daemon's account accessible to the scientific community and thus, symbolically at least, to a language of shared experience. It is this translation, as it were, which must be reversed if the description is to become truly representational.

[101] Cf. earlier quotation, p. 94, *supra*.

ered by Galileo. Kepler was fascinated by this phenomenon and already came up with an explanation for it in the *Conversation.*[102] This earlier explanation is elaborated in the "Geographical or, if You Prefer, Selenographical Appendix" to the *Dream*.

Galileo, we saw, had pointed out that these spots were cavities and sought to account for this phenomenon by focusing on one particularly big cavity which he considered analogous to the topography of Bohemia. In contrast, Kepler offers an explanation that accounts for the great number of cavities: In view of their perfectly round shape and their regular arrangement on the lunar surface, Kepler argued that they could not be attributed to a natural cause; he therefore suggested that these features had been constructed by the Endymionides, the rational inhabitants of the Moon, as fortified cities to shield them against the atrocity of the lunar climate and against enemies. He gives an elaborate description of the way the Endymionides measure out the space for their future cities by means of a primitive form of compasses:

> They drive a stake down in the center of the space to be fortified. To this stake they tie ropes which are either long or short depending on the size of the future town. ... With this rope fastened in this way, they move out to the future rampart's circumference, as defined by the ends of the ropes.[103]

> Palum defigunt in centro spacij muniendi; ab hoc palo nectunt funiculos, pro capacitate future oppidi longos an breves ... Cùm hoc sic defixo funiculo ad circumferentiam excurrunt futuri valli, quam fines designant funiculorum.[104]

Along with the fact that these lunar fortifications bear a resemblance to medieval cities, we should note Kepler's decision to interpret the crater as artificial rather than natural. His interpretation is based on two axioms listed in a note to the appendix. They state that "the mind is the source of orderliness" and that "[i]t therefore follows that those things which are out of order, insofar as they are out of order, are the result of the movement of the elements and the composition of matter."[105] On the other hand, "when things are in order, if the cause of the orderliness cannot be deduced from the motion of the elements or from the composition of matter, it is quite probably a cause possessing a mind."[106] The circular phenomena on the lunar surface, so Kepler argues, are too perfect in shape to spring from a natural cause; therefore, they "are artificial and produced by some architectural mind."[107] By attributing the phenomenon to the Endymionides, Kepler avoids having to explain a natural phenomenon that clearly distinguishes the lunar surface from the terrestrial one. Thus, he salvages the concept of comparability between the Earth and the Moon, just as Galileo did before him when he manipulated the visual evidence available to him.

---

[102] *Kepler's Conversation,* 88. See also, p. 51, *supra.*

[103] *Dream*, 151.

[104] *GW*, 368.

[105] *Dream*, 157. Adaptation mine. – "Mens sit author ordinis ... sequitur, ut ea, quae sunt inordinata, in quantum inordinata, sint à motu elementorum et necessitate materiae." (*GW*, 371)

[106] *Dream*, 159. – "Ea, quae sunt ordinata, si ordinis causa deduci nequit à motu elementorum aut à necessitate materiae, valde probabile est, esse ab alique causa mentis compote." (*GW*, 372)

[107] *Dream*, 160. – "ab aliqua mente architectrice." (*GW*, 373)

Moreover, that the two worlds are basically comparable is maintained through the principle of proportionality which, Kepler believes, not only links the two worlds, but pervades the entire universe. In the dream narrative, Kepler introduces it in connection with the average size of lunar creatures: since the mountains on the Moon are proportionally higher than those on Earth, Kepler assumed that lunar creatures were much larger than creatures on Earth.[108] "This proportion is applicable not only to their bodies in comparison with our terrestrial bodies but also to their functions, breathing, hunger, thirst, waking, sleep, work, and rest."[109] It links the topographies and forms of life on Moon and Earth both to their respective motions in space as well as to the phenomena apparent from their particular location in the cosmos:

> Between the very slow motion of the fixed stars for us and the brief periods of the individual planets down to the daily rotation of the earth there is a proportion which seemed to me to be also that of human life to the modest size of our bodies. For the moon, on the other hand, the fixed stars return more quickly than Saturn, whereas a day is thirty times longer than ours. Hence I thought that I should attribute a short life to the living creatures but enormous growth, so that nothing would attain a stable state and everything would perish in the midst of its development.
>
> Quae proportio motus fixarum lentissimi penes nos, ad restitutiones singulorem Planetarum, ipsiusque adeò Telluris diurnum, breves: ea mihi visa est esse proportio aetatis humanae ad molem corporum modicam. Ergò vicissim in Luna, cùm fixae revertantur celeriùs, quàm Saturnus, dies contrà sit trigecuplo longior nostra: brevem animantibus vitam, incrementa maxima tribuenda censui: ut nihil ad consistentiam perveniat, omnia in medijs intereant incrementis.[110]

This principle, which sets the different worlds in the universe in strict proportion to each other, represents the logical continuation of the principle of environmental adaptation mentioned earlier. Taken as the fundamental creative law of the entire cosmos, it not only vouches for a cosmos of aesthetic harmony, but constitutes the actual *raison d'être* of Kepler's lunar picture, for it offers a way out of the labyrinth of mirrors that forms the dream narrative by suggesting that the Daemon's description of the lunar world, if not literally true, is at least proportionally true.

Allegory, too, depends on the strict proportionality between the literal and the implied level of signification. In the passage quoted above, Kepler calls attention to the considerations that guided his imaginative creation of the lunar world. He thus again (though less explicitly than in the note quoted earlier)[111] sets his fictional creation parallel to God's actual creation of the cosmos, which it aims to represent. However, viewed within the framework of an allegory, this parallelism does not express an

[108] Kepler had already established that the lunar mountains are proportionally higher than the mountains on Earth in his *Optical Part of Astronomy*. In his *Conversation*, Kepler corrected this statement and joined with Galileo, who claimed that the mountains on the Moon are higher than those on Earth not merely in relation to their globes, but absolutely. This claim proved to be wrong. (*GW*, vol. 2, 128, and *Kepler's Conversation*, 31)

[109] "Nec tantùm corporum ea proportio ad nostra terrestria; sed etiam facultatum, respirationis, famis, sitis, vigiliarum, somni, laboris, quietis." (note 212)

[110] *Dream* and *GW*, note 213.

[111] Cf. pp. 94-95, *supra*.

equivalence, it merely sets the literal sense (Kepler's lunar fiction) in proportion to the implied allegorical sense (the true description of the lunar world as God created it). Kepler's allegorical narrative may not bear an intrinsic connection to the meaning it implies: since it is made up of images that he borrowed from a different context, its graphic presentation is largely arbitrary. Yet, as the commentary emphasizes, there is a logical connection between the two meanings; this connection gives the literal narrative the potential to point beyond itself to a reality for which there are neither words nor images.

To conclude this chapter, I would like to return briefly to an old dream that we mentioned long ago. Doing so will not only involve a look into the past, but may also inspire hope for a possible denouement of Kepler's allegory. In reference to the models for both the motif of the literary dream and its Icelandic setting, Kepler notes that the final part of Plutarch's *De facie* also relates a journey to an island that "a modern geographer would probably apply to the Azores, Greenland, and the territory of Labrador, regions situated around Iceland."[112] Kepler adds: "Every time I reread this book by Plutarch, I am exceedingly amazed and keep wondering by what chance it happened that our dreams and fables coincided so closely."[113] The part of Plutarch's book Kepler is referring to belongs to a separate narrative told by Sulla that concludes the dialogue. The correspondences between this narrative and Kepler's are indeed striking.

In introducing his narrative, Sulla assures his audience that he is "but the actor of the piece" since he heard the tale from a stranger, who claimed to have participated in the journey himself.[114] Sulla moreover points out that the author of the narrative started his account with a quotation from Homer, that is, another author. In a clear parallelism to the *Dream*, the narrative is thus also characterized by indirection.

Every thirty years, so the story goes, a Greek community that lives far out in the Northern ocean sends off a group of envoys to the island where Cronus, banished by Zeus, sleeps confined in a deep cave of rock. There they serve the sleeping god for thirty years, as is the custom, employing their time in sacrifices and celebrations or in the study of philosophy and astronomy. There are also spirits ("*daimones*") on the island who tend and serve Cronus and who have oracular power. They offer the men insight into the future in the form of prophecies; "but the prophecies that are greatest and of the greatest matters [the spirits] come down and report as dreams of Cronus, for all that Zeus premeditates Cronus sees in his dreams."[115]

This is the framing narrative. The tale within the tale consists of a mythical account of the Moon, in which the Moon is described as the place where the souls of those who lived morally good lives are temporarily conveyed. Sulla explains at the end that the stranger heard this account from "the chamberlains and servitors of Cronus", thus sug-

[112] For the original text, see pp. 80-81, *supra*.

[113] "Quem quidem Plutarchi librum quoties relego, toties impensè soleo mirari, quo casu factum sit, ut nostra nobis somnia seu fabulae tam accuratè congruerent." (note 2)

[114] *De facie*, 181. – "Ego enim jam deinceps actor sum". (*GW*, 404) This is Kepler's own Latin translation of Plutarch's text, which he appended to the *Dream*.

[115] *De facie*, 189. Adaptation mine. – "maxima verò et de rebus maximis tamquam somnia Saturni ['*tou Kronou*' in the original], enunciare descendentes; Quicquid enim Jupiter meditetur, id Saturnum somniare". (*GW*, 405)

gesting that it represents one of Cronus' dreams and that we are dealing once more with a lunar dream.[116]

Yet here too the situation is ambiguous. For even while Sulla explicitly attributes the account of the Moon to Cronus' servants, there is a curious passage immediately preceding the account which seems to contradict this attribution: The passage relates how the traveler left the island and "discovered certain sacred parchments that had been secretly spirited off to safety when the earlier city was being destroyed and had lain unnoticed in the ground for a long time."[117] Without any transition, the stranger then introduces the lunar myth. This creates the impression that the myth was contained in the parchments, especially since their mention is otherwise unmotivated.

Besides being a prophecy about the future, the lunar myth thus emerges as a piece of ancient history. As such it is also a quotation, for the reference to the sacred parchments clearly evokes the passage at the beginning of Plato's *Timaeus*, where Critias introduces an old story of a war fought by an earlier Athens against Atlantis (another island!). The story has a complex origin: Critias says he had it from his grandfather, who in turn had it from Solon, who heard it on his travels in Egypt. There it had been preserved in a written record kept in Egyptian temples and thus survived the catastrophic events that led to the complete destruction and disappearance of both the city of Athens and the island Atlantis. The myth of Atlantis – which Kepler mentions together with Plutarch's *De facie* and Cicero's *Somnium Scipionis* in the same note – thus includes a story of transmission as complex as that of Sulla's narrative and Kepler's own. The fact that both narratives allude to this myth emphasizes the strong ties that bind these texts to a tradition of myth-making.

The different levels of narration, the multiplicity of textual references, the island, the years spent serving a god, the astronomy, the daemons, the opposition between the stranger's experience of a physical journey and the dream visions or imaginary journeys arising from Cronus' confinement, the combination of sitting in the dark and thus not seeing (Cronus in his cave) and 'seeing' in a dream, finally, the dream itself, the dream about the Moon that is – all this we meet again in adapted form in Kepler's *Dream*. And on top of all this, the ambiguous attribution of the lunar myth in Sulla's narrative reflects the questionable identity of Kepler's account with singular aptness, presenting it as both regurgitation and prophecy: As a testimony of the past, Sulla's lunar myth bears testimony to the dependence of Kepler's fiction on the fictions of the

---

116 *De facie*, 223. – "illos Saturni cubicularios et ministros". (*GW*, 409) There is an interesting parallelism here to the *Tabula of Cebes*, for the painting on which the allegorical narrative is based is located in a temple consecrated to Cronus. Moreover, the old man who explains the painting to the first narrator of the book (again, there are several levels of narration!) claims to have been given the account by a venerable stranger, who dedicated the temple and the painting to Cronus. Since the date of composition of Cebes' *Tabula* is uncertain (though critics generally ascribe it to the first century BC), it is not clear whether Plutarch's text was the source here or vice versa. Yet since until the end of the eighteenth century the *Tabula* was generally believed to have been written by Cebes of Thebes, a contemporary of Socrates, Kepler would probably have taken the *Tabula* for the older text. Again, the provenance of the narrative material becomes more and more complex. For an account of the theories around the dates of composition of the *Tabula*, see the introduction by Fitzgerald and White to their edition of the text, 1-7.

117 "quia sacras paginas, tùm quando prior urbs eversa fuit, clam surreptas, multoque tempore latitantes, subque terram defossas, ipse vestigando invenit." (*GW*, 406)

past. As Cronus' dream, the account of the Moon is a prophecy about the future since Cronus sees in his dreams all that Zeus premeditates. Cronus thus sees the image of that which has no image because it has not yet come to pass. In figurative terms, the phrase could simply mean that what Zeus premeditates will be seen and lived in time, that is, *chronos*. Similarly, the work of astronomers may bear fruit in time so that Kepler's lunar allegory can be exchanged in time for a literal account.

4

Worlds of Words: Cyrano de Bergerac's *Lune* and *Soleil*

In the light of the preceding observations on Galileo and Kepler and the implicit argument regarding the impossibility of envisioning alterity that runs through their descriptions of the Moon, Cyrano de Bergerac's *Empires et Estats de la Lune* and the *Empires et Estats du Soleil* read both like a faithful illustration of this argument and its *reductio ad absurdum*. In other words, the two-part novel reads like Ariosto's lunar narrative set in a scientific context. For this reason, I would first like to look at Cyrano's text in isolation, focusing on those aspects that build on what has been said so far. At the same time, his novel also offers a transition since the argument of another world is extended here to that of the plurality of other worlds, which will be the topic of the next chapter. There the discussion will mainly focus on Fontenelle's *Entretiens sur la pluralité des mondes*, but will include illustrations from several other texts that introduce the concept of the plurality of worlds, among them Cyrano's *Lune*.

The theory of the plurality of worlds is in many respects a direct consequence of Copernican theory, even if Copernicus himself would have been loath to admit this.[1]

[1] In his influential work *The Great Chain of Being: A Study of the History of an Idea* (1933) (Cambridge: Harvard University Press, 1976), Arthur Lovejoy denies the Copernican Revolution any significant role in the development of the concept of plural worlds, arguing instead that the concept derived from the Neoplatonic *principle of plenitude*. He illustrates this thesis with a history in which he refers to speculations around the existence of plural worlds that were unrelated or at least not directly related to Copernican theory. Karl Guthke disagrees with this position and points out that it was through Copernican theory that the ancient idea of plural worlds acquired a higher degree of plausibility. Among other things, he criticizes Lovejoy for not clearly distinguishing between the idea of a plurality of *kosmoi* (which goes back to Aristotle) and a plurality of worlds within a single cosmos. He also points out that the principle of plenitude, symbolized through Homer's golden chain, the Great Chain of Being, represented a strictly vertical order which clearly adhered to the geocentric cosmos and the theological hierarchy to which the principle was applied. Such an order could not span the ubiquity of life in a post-Copernican universe of innumerable planetary systems. Cf. Karl Guthke, *Der Mythos der Neuzeit: Das Thema der Mehrheit der Welten in der Literatur- und Geistesgeschichte von der kopernikanischen Wende bis zur Science Fiction* (Berne and Munich: Franke, 1983). While in general I agree with Guthke's argument, I think that his radical defense of Copernicanism with regard to its influence on the theory of the plurality of worlds is just as unnecessarily one-sided as the view he opposes. As Lovejoy's examples and many of Guthke's own examples show, the principle of plenitude continued to play an important role in speculations around the plurality of worlds. It would

Once the Earth was categorized among the planets that circle the Sun, the issue of the existence of other correspondences between the Earth and the rest of the planets followed logically. With the first telescope observations suggesting that the Moon was essentially earthlike in nature, further questions arose regarding the existence of life on the Moon and, more particularly, forms of rational life. These speculations were then extended by analogy to the other planets of our solar system. While a more conservative Copernican like Kepler still restricted his speculations on the plurality of worlds to the planets circling our Sun, in the second half of the seventeenth century it became increasingly common to interpret the fixed stars as other suns, each with its own planetary system. The theory of the plurality of worlds thus developed into a theory of the infinity of worlds, and Bruno's heresy gradually turned into a form of liberalized theology, which saw in the existence of innumerable worlds and the ubiquity of life in the universe both a necessary consequence of and condition for the perfection of divine creation.[2]

In the *Lune* and the *Soleil*, Cyrano (through his anagrammatical namesake Dyrcona) explicitly argues in favor of a plurality, even an infinity of inhabited worlds. Cyrano bases his argument on the scientific theories of his time, particularly on Descartes' atomistic cosmology of vortices, which postulates the indeterminate extension of matter. Unlike Descartes, Cyrano explicitly assumes the existence of as many Copernican planetary systems as there are suns. Though presented in the form of a fictitious narrative, the theory Cyrano proposes in his novels thus combines philosophical speculation with a complete physical system. At this point, however, I am not yet interested in Cyrano's argument in favor of the theory of plural worlds; this theory is mainly presented in the form of speculative conversations Dyrcona has first with the Viceroy of Canada before his trip to the Moon and later with the philosophers on the Moon. Rather, I want to focus on the descriptions of the world on the Moon and the Sun and Dyrcona's encounters with their inhabitants. These descriptions take up the greater part of Cyrano's two narratives and comment in an exemplary way on the nature of imaginative creations.

Yet before we do so, we should once again take a close look at the way the narrative defines the rhetoric of its contents: Cyrano's descriptions of the lunar and the solar world are based on the fiction of two actual voyages into space that the first-person narrator Dyrcona undertakes in close succession. This, at least, is what the narrative relates on the surface: Though these voyages are described in some technical detail, the framing narrative as well as the accounts of Dyrcona's sojourn on the Moon and the Sun abound with references to sleep and dreams so as to leave open the possibility that the journey into space only happened in a dream. There is thus a metanarrative dimension to Cyrano's text that closely resembles that of Kepler's *Somnium*. In reference to

---

therefore be more accurate to say that in such speculations, this principle and arguments derived from it often performed an important function as a priori philosophical principles that justified the concept of plural worlds *on the theoretical foundation of Copernicanism*. We will consider some instances of this in the next chapter. For a more balanced history of the theory of the plurality of worlds, see Stephen J. Dick, *Plurality of Worlds: The Origins of the Extraterrestrial Life Debate from Democritus to Kant* (Cambridge: Cambridge University Press, 1982).

2 Guthke, *Mythos*, 122.

Dyrcona's first, failed attempt to travel to the Moon, which takes him to Canada instead,[3] Timothy Reiss has observed that

> the narrator will imply that his first journey is the result of a dream, and even perhaps that it *is* a dream: 'In order to make an end of it, I shut myself away in a comparatively isolated country house where, having gratified my daydreams with some practical measures appropriate to my design, this is how *I gave myself to* the sky'."[4]

There is another extended reference to dreams shortly before Dyrcona's imprisonment and his subsequent journey to the Sun: One night, while visiting with two friends near Toulouse, Dyrcona and one of the friends are visited by dreams very similar in content, involving catastrophe as well as a flight to the Sun. Dyrcona attributes his dream to his present state of mind and their occupations of the previous day. He is confirmed in this opinion by the other friend, Cussan, who is amazed at how much their dreams

> are a hotch-potch of everything we have been thinking about when we were awake, a monstrous chimera, an assembly of confused elements, which our fancy – no longer guided by our reason when we are asleep – presents to us in complete disorder. Yet, none the less, we believe that, by twisting them about, we can squeeze the true sense out of them and extract knowledge of the future from dreams, as we can from oracles.[5]

> c'est un pot pourry de toutes les choses à quoy nous avons pensé en veillant, une monstreuse chimère, un assemblage d'espèces confuses, que la fantaisie qui dans le sommeil n'est plus guidée par la Raison nous présente sans ordre, et dont toutefois en les tordant nous croyons épreindre le vray sens, et tirer des Songes comme des Oracles une science de l'avenir;[6]

Cussan here identifies their dreams as a random recycling of past thoughts, random because these thoughts resurface *in complete disorder*. He then contrasts this definition of dreams with the common belief in the prophetic quality of dreams which holds that by 'twisting them about', i.e., by *reordering* them, we 'can extract knowledge of the future from dreams'. In the light of Cussan's analysis of the two dreams, we are forced to dismiss this belief as nonsense, for by reordering something that has lost its previous order, we either get back the old thing in its old order or a transformed version of it, the same elements in yet another order; we do not get something new. Accordingly, by

---

3 This journey is only partly a failure, however, since Dyrcona takes it as proof that the Earth turns around itself: Given that his rise and descent occurred in a line perpendicular to the ground, the Earth must have turned under him while he was up in the air.

4 *Discourse*, 236. Italics T.R.

5 *Other Worlds: The Comical History of the States and Empires of the Moon and the Sun*, ed. and trans. Geoffrey Strachan (London: Oxford University Press, 1965), 116. Quotations in English are taken from this edition, henceforth cited as *Other Worlds*.

6 *Les Estats du Soleil*, in *Les oeuvres libertines de Cyrano de Bergerac* (1921), ed. Frédéric Lachèvre (Geneva: Slatkine Reprints, 1968), vol. 1, 107. All quotations in French are taken from this edition, henceforth cited as *Lune* and *Soleil*, respectively. I have removed the editor's italics, which were used to mark different manuscripts.

reordering confused elements of the past, we either get back the past, or a skewed version of the past, but hardly the future, which is beyond the experience available at the time. What then are we to do with the fact that both dreams, but especially Dyrcona's, offer an accurate summary of the events that follow in the narrative: Dyrcona's imprisonment and his successful escape from the prison by means of a flying machine which takes him to the Sun? Are we to read the ensuing narrative as the confirmation of prophetic dreams by history or just as a detailed rendering of the story first given in the form of a summary, a *rubrum*?

While Cussan's comment on the nature of their dreams dismisses the first interpretation, the second one seems to be corroborated, first of all, by Dyrcona's remark that this kind of dream has always been common with him ("Je n'ay guère eu que des songes semblables à celuy-là depuis que je me connois")[7]. It is also supported by the fact that the entire account of Dyrcona's adventures on the Sun is full of references to his falling asleep and dreaming or to moments when some strange occurrence prevents him from sleeping.[8] The most notable example is the episode at the Lake of Sleep, where Dyrcona falls asleep again and dreams one of the most erudite and best conceived dreams ever ("un songe le plus sçavant et le mieux conceu du monde")[9]. The dream is interrupted by his guide Campanella, but Dyrcona promises to give his readers an account of it later on. This account is never given, or at least not explicitly. However, the episode with the couple from the province of lovers that follows has a strikingly dreamlike quality to it; and shortly after this episode, the narrative suddenly breaks off, just as both the dream at the lake and the dream in Toulouse were interrupted.[10] Finally, just before the *Soleil* comes to an abrupt end, Campanella anticipates Descartes' arrival long before Dyrcona can see him. Dyrcona concludes from this that Campanella must have seen the philosopher in a dream. Though Campanella will explain this act of clairvoyance scientifically by means of a theory which closely resembles Lucretius' theory of *simulacra*,[11] his first answer to Dyrcona's assumption is ambiguous: "'If you call a 'dream' ... what you can see in your mind as clearly as you see with your eyes by day, when the sun is out, then I admit this.'"[12]

---

7 *Soleil*, 106.

8 Every major episode on the Sun is introduced by some such reference. Accordingly, as soon as Dyrcona has arrived on the Sun, he is overcome by great tiredness and falls asleep. When he wakes up, he encounters the nightingale, the metamorphosed king and his nation. Later he follows the nightingale, loses sight of it, and is about to abandon himself to sleep again when he suddenly sees the Phoenix. The encounter with the mythical bird leads to the history of the birds. After having continued on his journey, Dyrcona takes a rest and is again about to fall asleep when he hears voices. The episode with the talking trees and the fight between the remora and the salamander follows. It leads to the encounter with Campanella, which in turn brings about the episode at the Lake of Sleep.

9 *Soleil*, 187.

10 Dyrcona's dream in Toulouse was interrupted when he suddenly fell out of his bed: "I would doubtless have observed many other things there, but with my flying I had been tossing about, which brought me so near to the edge of the bed that I fell on to the floor beside it, with my belly completely bare on the plaster and my eyes wide open." (115-16)

11 Cf. p. 56, *supra*.

12 *Other Worlds*, 232. – "Si vous appellez songe ... ce que vostre âme peut voir avec autant de certitude que vos yeux le jour quand il luit, je le confesse." (*Soleil*, 198)

Both the *Lune* and the *Soleil*, but more notably the latter, thus constantly suggest the possibility that what they relate is only a dream – a dream, moreover, that is a hotch-potch (*pot pourry*) of thoughts and ideas of the past. The word *pot pourry*, which Cussan uses to characterize the dreams of his friends, appears just a few pages earlier when Dyrcona refers to his account of his journey to the Moon.[13] It is worth looking at this passage in context since the narrative here offers a critique of its contents that closely resembles Cussan's evaluation of the two dreams: At the beginning of the *Soleil*, Dyrcona relates how his friends persuade him to write a book about his extraordinary adventures on the Moon. Once published, the *Lune* stirs up strong reactions ranging from initial storms of enthusiasm to charges against Dyrcona that accuse him of being a black magician. He comments on this with a humorous recantation:

> And thus behold the medal reversed and see how they vie with one another in recanting! The work they had previously made so much of is now nothing but a hotch-potch of ridiculous stories, a heap of disjointed fragments, a collection of bedtime fairy tales for children;[14]

> Voilà donc la médaille renversée, c'est à qui chantera la palinodie. L'ouvrage dont ils avoient fait tant de cas, n'est plus qu'un pot pourry de contes ridicules, un amas de lambeaux décousus, un répertoire de Peau-d'Asne à bercer les enfans;[15]

Dyrcona's recantation is obviously ironic. This is evident both from its tone as well as from the fact that it ironically echoes the scene on the Moon when Dyrcona is condemned to a public recantation of his claim that the 'moon' (our Earth) is a world too, a recantation that in its own turn re-enacts the trial by the Inquisition against Galileo.[16] Yet it is the phrasing of this recantation that is interesting: In analogy to Cussan's later characterization of their dreams as a *pot pourry* of thoughts of the past, the *Lune* as a whole (and, by implication, the *Soleil*) is dismissed as a *pot pourry* of ridiculous tales and disconnected fragments, a repertoire of fairy tales, in short, a wild mixture of old stuff ripped out of context. Indeed, I cannot think of a more fitting description of Cyrano's text! The account of Dyrcona's two journeys is a curious mixture of quotations, parodies, emblems, metaphors and allegory. Before we take a closer look at this *pot pourry*, there is another phrase in this passage that merits close attention: Dyrcona refers to his narrative as *un répertoire de Peau-d'Asne à bercer les enfans*. This phrase is used with more purpose than might first appear, as it includes both a criticism of the text's fictional identity and its apology in emblematic form.

At the time Cyrano wrote his novel, *Peau-d'Asne* (or *Peau d'âne* in modernized spelling) was already used as a generic term for a fairy tale. More specifically, however, the name referred to an old folk-tale in which a princess, to escape her father's incestuous love, hides in an *ass's skin* and in this humble disguise wins the love of a

---

13 *Soleil*, 101.

14 *Other Worlds*, 109. A little later, on realizing the danger of being condemned to death for his book, Dyrcona comments in a similar vein: "For I should be doubly vexed to die for something I can scarcely believe myself." (*Ibid.*, 111) The French is even stronger: "une chose à laquelle je ne crois guères". (*Lune*, 103)

15 *Soleil*, 101-2.

16 For the inversion of Galileo's recantation, see Guthke, *Mythos*, 178.

prince, who sees through her disguise. The tale thus involves both fake appearance and the ability to recognize the true identity that lies behind this appearance. If we apply it to the *Lune*, Dyrcona's statement could be taken to suggest that even while the description of his journey appears like a repertoire of fairy tales, this is only the superficial appearance of the text, behind which lies hidden its true and noble identity. The theme of penetrating a hidden truth recalls Galileo's sexual metaphor in the *Sidereus nuncius*, the astronomer's gaze that penetrates to the Moon's most secret parts: again, the male knower who sees through fake appearances.[17]

In another close correspondence to Kepler's *Somnium*, Cyrano thus both deconstructs the outward presentation of his argument and rehabilitates it on a deeper level. As in the case of the *Somnium*, the distinction between a superficial appearance and a deeper truth refers to the terms in which the lunar and the solar world are described in the text; these terms dramatize in a powerful way the impossibility of knowing the true appearance of these worlds and of the limited number of ways in which their fictional substitutes in the imagination can be created. Kepler, I argued, presented his description of the lunar world as an allegorical substitute for its true description and justified the use of allegory through the principle of proportionality. The lunar world which the Daemon describes is thus still related in appearance to the true image of the Moon implied by the allegory. In contrast, Cyrano's fictions of the lunar and solar world do not in any way pretend to hold such a relation to the worlds they signify; as representations, they are completely fictitious. The allegories, metaphors, emblems, quotations and parodies which his fictions largely consist of do not refer to any reality beyond their own; Cyrano's fiction offers a vivid illustration of the limitations on the imagination's creativity, one closely resembling Ariosto's anti-allegory of allegorical representation.

Yet parallel to this illustration, the *asinine* appearance of Cyrano's account has its justification in the theory which holds that different appearances merely result from different arrangements of the particles contained in the same overall substance. This theory is explained in terms most suitable to our context by Gonsales, a character Dyrcona meets on the Moon: "From a serious investigation of matter, therefore, you would discover that it is all one, but that, like an excellent actor [*une excellente comédienne*], it plays all kinds of roles in this life, in all kinds of disguises."[18] The theory is, moreover, acted out in an episode on the Sun that we will turn to later. In short, appearances do not matter: ass and princess are essentially the same.

Nevertheless, the play with outward appearances is one of the most striking characteristics of the text: Dyrcona's descriptions of the worlds on the Moon and the Sun abound in characters, places and whole episodes that are taken from literary texts both ancient and contemporary. Thus Dyrcona, on landing on the Moon, finds himself in the Terrestrial Paradise. There he meets Elija, who gives him a guided tour of the place and its history that is a travesty of the story told in the Genesis. Later, Dyrcona meets Domingo Gonsales, the protagonist of Francis Godwin's *The Man in the Moone*, whose flight to the Moon with the help of "Gansas" (some sort of wild geese) had

---

[17] Cf. p. 44, *supra*, and Timothy Reiss, "Espaces."

[18] *Other Worlds*, 45. Addition mine. – "et ainsy à pénétrer sérieusement la matière, vous trouverès qu'elle n'est qu'une, qui comme une excellente comédienne, joue icy bas touttes sortes de personnages sous touttes sortes d'habits". (*Lune*, 46)

become famous through many editions and various translations that the book appeared in.[19] Cyrano moreover adopted ideas and entire episodes from works by Charles Sorel, particularly the popular *Histoire comique de Francion*, where the protagonist visits the Moon and the Sun in a dream.

Of the literary references in the *Soleil*, some are similar to the ones just mentioned in so far as they refer to a specific source: The description of the Kingdoms of Lovers and of Truth, clearly based on Campanella's *Città del Sole*, is a good example in this respect, especially since this episode takes place in Campanella's own presence. Another example is the implicit reference to Dante's *Commedia*, which functions as a shadow text to Dyrcona's second journey: In an obvious parody of Dante's guided tour through the realms of the *oltremondo* at the hands of Virgil and Beatrice, Campanella here serves as Dyrcona's guide around the world in the Sun. Like Dante, Dyrcona is allowed to see and learn about things that no human being before him ever knew while alive; he is therefore entrusted with the task of instructing his fellow human beings on Earth about what he learned on his journey.[20] Other references are of a more generic type: Dyrcona's meeting with the Phoenix and the talking trees, for instance, take up two classical literary *topoi* rather than referring to a specific literary work.[21]

It would be interesting to undertake a detailed analysis of the specific use Cyrano makes of his literary borrowings by recontextualizing and changing them, but this is not our purpose here. Instead, I want to focus on the references as such and on the way the text displays its intertextual nature. For in contrast to the *Somnium*, where, as we have seen, sources are given in a separate commentary, Cyrano's text is openly intertextual: Most of the quotations are either obvious or explicitly attributed to their sources. The worlds in the Moon and the Sun are clearly literary worlds. Nevertheless, the narrator does not acknowledge the fact that the episodes and ideas quoted are fictional borrowings from other texts, but presents them either as empirical confirmations or corrected versions (based on the new empirical evidence that Dyrcona's journey supplied) of the events related in these texts.

A closer look at the quotations mentioned above will illustrate this point. On being told that among the inhabitants of the Moon poetry is used as money, Dyrcona thinks to himself: "'Well to be sure, ... this is just the currency which Sorel makes Hortensius use in *Francion*, I remember. Doubtless this is where he got it from'."[22] What Francion dreamed is thus confirmed by Dyrcona's trip (which, we said, may have also only been

---

19 The story was first published in 1638. Ten years later the French translation appeared, and it is this translation that Cyrano must have used since he did not know English. Cf. Lachèvre's comment in *Lune*, 44, note 2.

20 Campanella makes this clear to Dyrcona in the following comment: "'You are lucky indeed,' said Campanella, 'to be seeing all the wonders of this world before you die, and it is fortunate for the inhabitants of your globe that it will have carried a man who can teach them about the marvels of the sun, since, but for you, they would have run the risk of remaining in gross ignorance for ever and enjoying a hundred good things, without knowing their source. For you cannot imagine the riches that the sun bestows upon all your little globes'." (*Other Worlds*, 215)

21 With regard to the episode of the talking trees, it should be noted, however, that it not only quotes a classical *topos*, but also the specifically Dantean episode of the suicides metamorphosed into trees (*Inferno*, XIII). In this way, the parallel to the *Commedia* is more firmly established.

22 *Other Worlds*, 42. – "'Ha! vrayment, ... voilà justement la monnoye dont Sorel faict servir Hortensius dans *Francion*, je m'en souviens. C'est là, sans doubte, qu'il l'a dérobé". (*Lune*, 43)

a dream). However, Dyrcona's lunar encounter with Domingo Gonsales at once confirms and disproves the fiction in *The Man in the Moone*, from which Gonsales originates: The confirmation implied by the Spaniard's presence on the Moon is offset by the fact that Dyrcona's description of the lunar world differs in most respects from that given in Wilkins' novel. Dyrcona's encounters with the Phoenix and the talking trees offer an analogous situation: From the Phoenix, Dyrcona learns that rather than burn in his nest to give birth to a new bird (the traditional myth), the Phoenix flies to the Sun to join the community of birds there. Later, the talking trees insist that the reason and the soul inside them are their own; they are not human beings metamorphosed into trees as has been claimed time and again in Western epics.

Finally, there is the story of Elija: The prophet's mysterious ascension to heaven is explained scientifically in an extended version of the original Biblical story. In the *Lune*, Elija builds a "chariot" that takes him to the Moon by means of magnetic attraction. In addition to this elaboration of the original story, the *Lune* also offers a sequel about Elija's subsequent life in another world on the Moon. Rather than take Elija to a new world, however, the journey takes him into the past (that is, to the Terrestrial Paradise) so that the sequel is turned into a reappraisal and travesty of the old story in the Genesis. The differences between the traditional version of this story and Elija's own do not lie in any radical deviations from the old story, but in the way Elija represents and embellishes this story.

In spite of all the adaptations and elaborations performed on such literary fictions, they are thus in a curious way taken literally: It is, after all, still Gonsales, the Phoenix, the Tree of Life, etc. that Dyrcona meets on the Moon and the Sun. Detached from their original context, they materialize into the reality of another world into which they are bound by new relations (both causal and narrative) only to be recontextualized in an amalgam of literary fragments that remain clearly identifiable.

As in the literary adaptations that largely constitute Cyrano's lunar and solar worlds, the literalism characteristic of these worlds also manifests itself in episodes where metaphorical expressions or simple figurative terms already standardized in common linguistic usage are used literally, with a meaning opposed to the habitual figurative one. Lucia Cenerini gives some good examples of this practice: Noah's daughter Achab, called a *lunatic*, abandons her family on the ark and embarks on an adventurous exploration of the Moon. Dyrcona satisfies his empty stomach with a flock of larks falling roasted from the sky, which corresponds to the related figurative expression *il attend que les alouettes lui tombent toutes rôties*. It also, of course, enacts the stereotypical image of paradise that figures on the Moon in the double sense of generic and Biblical paradise. Moreover, the standard expression *recueillir son dernier soupir* is dramatized in a scene where the best friend of the philosopher who decided to end his life "ne destache point ses lèvres de celles de son Amant, qu'il ne le sente expiré;" in a cannibalistic rite the corpse of the deceased is then eaten by his dearest friends, an act that corresponds to the expression *je t'aime tant que je te mangerais*.[23]

---

23 "Metafora in utopia" in *Il romanzo barocco tra Italia e Francia* (Rome: Bulzoni, 1980), 231 and 234.

The most significant episode in this respect is the one in which Dyrcona's guide on the Moon, the so-called Demon of Socrates,[24] pays their bill in the pub with poetry. He explains to Dyrcona that on the Moon poetry serves as money, a custom that literalizes the expression *se payer de mots*.[25] The organization of this monetary system is illustrated in the following passage:

> When some have been composed, the author brings them to the *Court of the Mint*, where the Jury of Poets of the Realm hold their sessions. There these Officers of Versification [sic! – les *vérificateurs*] put the pieces to the test and if they are juged to be of good *alloy*, they are *taxed* not according to their weight but according to their wit, ... Thus when someone starves to death it is never anyone but a blockhead and witty people always live off the fat of the land.[26]

> Quand on en a composé, l'Autheur les porte à la *Cour des Monnoyes* où les Poëtes jurés du Royaume font leur résidence. Là, les *vérificateurs* officiers mettent les pièces à *l'espreuve*; et si elles sont jugées de bon *aloy*, on les *taxe*, non pas selon leur poids, mais selon leur pointe, et de cette sorte quand quelqu'un meurt de faim, ce n'est jamais qu'un bufle, et les personnes d'esprit font tousjours grand'chère.[27]

The semantic field associated with money is here directly applied to poetry so as to emphasize the equation between 'regular' money and poetry or, more generally, words. If words are money, they also become the equivalent of what money buys: the things they name. Accordingly, '*les personnes d'esprit font tousjours grand'chère*'. Poetry as money is thus the ultimate reification of the word.

So much for the *Lune*. Yet the literalization of figurative language is also effective in the *Soleil*: The episode around the suitors' hyperbolic love-talk in the Kingdom of Lovers provides both a further example of literalized metaphors as well as a commentary on the technique of literalization in general. This episode does not constitute one of Dyrcona's immediate experiences, but is part of a narrative of a woman Dyrcona meets on his tour of the Sun. We will see that the indirection is important since the episode concludes with an evaluation of the literalizing practice; as a result, the episode and its evaluation offer a *mise en abîme* and a metatextual evaluation of the primary narrative.

The woman tells Dyrcona that when she first came to live in the Kingdom of Lovers the misery and chaos she caused among her admirers made her so unhappy that she decided to escape. Yet her suitors warned her that, because of the many tears they had shed out of love for her, her house was besieged all round by water. Hard pressed as to what vessel she could use to cross this ocean, she listened to the advice of a jealous lover who suggested that she embark on the journey in her heart. He assured her that it would easily hold her since it already held so many others: "I should therefore set sail

---

[24] This is again a reference to another text, though one that is no longer evident to most modern readers. Lachèvre explains that this character was taken from the autobiography of Cyrano's friend Tristan l'Hermite. (*Lune*, 34, note 1) The Daemon may, of course, also be a reference to Kepler's Daemon.

[25] "Metafora in utopia," 231.

[26] *Other Worlds*, 42-43. Emphasis and correction mine.

[27] *Lune*, 43. Emphasis mine.

upon the sea of his tears; the blindfold of his love would serve me as a sail and the favourable winds of his sighs, despite the storming of his rivals, would bring me safely to port."[28] The woman is caught in the act of executing this advice (and herself) by another one of her admirers and is rescued at the last moment. The jealous lover is brought before the court, which forever bans him from the Kingdom of Lovers and forbids him on penalty of death ever to make use of hyperbole again.

The judges expose the lover's hyperbolic phrases as empty rhetoric, but the woman to whom they were addressed understood them literally.[29] Originally a native of the Kingdom of Truth, she is incapable of interpreting language in any other way and does not know what to make of the rhetoric used by her admirers or, for that matter, of rhetoric in general. In consequence, she misunderstands. In view of the clear similarity between the literalization of metaphors in the lovers' hyperbolic speeches and the literalized metaphors and figurative phrases in the account of Dyrcona's journeys, it seems justified to read the judges' critical evaluation of the former as a commentary on Dyrcona's narrative at large: Even while the literalization of metaphors contributes significantly to the construction of episodes and characters in the narrative, taking the result of this practice literally would be tantamount to misunderstanding it. The defiguring literalization of metaphors as we find it in Dyrcona's narrative paradoxically constitutes the ultimate figuration of language, its metamorphosis from letter into thing. Yet with regard to interpretation, it would be a fatal mistake to refer the text to any reality other than a textual one, as even its ultimate figuration remains a figure of speech. Cyrano's text thus constantly demonstrates how the fiction of the lunar and solar worlds that it presents is figured. It shows how the imagination visualizes these worlds through a playful reemployment of literary figurations and, more generally, of the figurations of language.

The woman's gullibility has its precedent in the story of the deceived nightingale, which Dyrcona hears from the king of a solar nation. This episode offers a paradigmatic illustration of the figuration of the solar world in visual images: Dyrcona meets with the king and his nation on awaking at the foot of a tree all made of gold and adorned with emerald leaves, buds of pearls and diamond flowers. Sitting on this tree is a nightingale engaged in singing the most exquisite melodies. Dyrcona then distinguishes a pomegranate of extraordinary beauty hanging on the same tree. In a spectacular metamorphosis, this fruit gradually humanizes into a dwarf and falls onto the ground next to him. The dwarf introduces himself to the amazed spectator: He is the king of the nation contained in the tree. In another metamorphosis, the tree suddenly disintegrates into a great number of little men. Only the nightingale remains the same and continues to pipe its melancholy love songs. When Dyrcona asks the king why the

---

28 *Other Worlds*, 230. – "Que je partisse donc sur la mer de ses larmes, que le bandeau de son amour me serviroit de voile, et que le vent favorable de ses soûpirs, malgré la tempeste de ses rivaux, me pousseroit à bon port." (*Soleil*, 197)

29 The practice of literalizing metaphors that is ironized here may be traced to a literary fashion conspicuously displayed in the work of some lyric poets greatly popular all over Europe in the early sixteenth century, notably the Spaniard Benedetto Carteo and the Italians Antonio Tebaldeo and Serafino Ciminelli d'Aquila. According to A. J. Smith, "conceited hyperbole is the life of the poets' style. ... The conceit has been worked up into a consistent figurative action." (*Metaphysical Wit*, 20-21)

nightingale has not changed its form, the king replies that it could not have done so: "it is a true bird, which is no more than what it appears to you to be."[30] The king then offers to tell Dyrcona the bird's story.

To facilitate their conversation, he and his subjects undergo another metamorphosis and turn into a midsized man. In this form, the king explains to Dyrcona that he and his people are natives of the light region of the Sun and that they spend most of their time traveling through the vast solar regions to study the diverse customs of their inhabitants, their climates and many other things. In their original form, the king and his nation are so subtle that they would have to travel with too much speed; they would, moreover, be imperceptible to Dyrcona's senses ("qui ne peut tomber sous tes sens")[31]. This is why, when they last set out on their journey, they metamorphosed into birds, the king adopting the form of a nightingale, his subjects that of eagles. In this configuration, they met with the true nightingale, who was shocked to see the other 'nightingale' in the power of eagles, "for it could not do otherwise than take us for our outward shapes".[32] Enchanted by its beautiful songs of love and sorrow, the king fostered the bird's deception for some time longer, but when this threatened to lead to the bird's self-destruction, he tried to break the illusion. His words and protestations were of no avail, however, so that he finally offered to prove his point by their metamorphosing before the bird's eyes into whatever form it chose. This is what they were engaged in when they met Dyrcona.

As with the woman from the Kingdom of Truth, the 'true' bird thus cannot help but take what it perceives literally; it judges by appearances alone. The point is, of course, that Dyrcona does the same, since to him, too, the king and his people manifest themselves only metaphorically through various metamorphoses. As the king explains, these metamorphoses are the product of their powerful imagination, which allows them to assume whatever shape they please:

> You must know that being inhabitants of the pellucid part [*la partie claire*] of this great world, where the principle of matter is to be in action, we are bound to have much more active imaginations than those of the opaque regions, and bodies composed of a much more fluid substance [*la substance du corps aussi beaucoup plus déliée*].
>
> Once one supposes this, it inevitably follows that our imagination, encountering no resistance in the matter of which we are composed, can arrange it at will, and, having become the complete master of our bodies, manoeuvres them, by moving all their particles, into the arrangements necessary to constitute the objects it has pictured in miniature. ... You men cannot do such things on account of the heaviness of your substance and the frigidity of your imagination.[33]

> Il faut que tu sçaches qu'estant nés habitans de la partie claire de ce grand Monde où le principe de la matière est d'estre en action, nous devons avoir l'imagination beaucoup plus active que ceux des régions opaques, et la substance du corps aussi beaucoup plus déliée. Or cela supposé, il est infaillible que nostre imagination ne rencontrant aucun obstacle dans la matière que nous compose, elle l'arrange

[30] *Other Worlds*, 158. – "C'est un véritable oiseau qui n'est que ce qu'il vous paroist." (*Soleil*, 140)

[31] *Soleil*, 142.

[32] *Other Worlds*, 161. – "car il ne nous pouvoit prendre pour tels qu'il nous voyoit". (*Soleil*, 142)

[33] *Other Worlds*, 164-65. Additions mine.

comme elle veut, et, devenuë maistresse de toute nostre masse, elle la vait passer, en remuant toutes ses particules, dans l'ordre nécessaire à constituer en grand cette chose qu'elle avoit formée en petit. ... Vous autres Hommes ne pouvez pas les mesmes choses à cause de la pesanteur de vostre masse, et de la froideur de vostre imagination.[34]

Although, according to this atomistic theory, the metamorphosed forms in which the king and his people manifest themselves to Dyrcona are only recompositions of the same matter, they would not be perceptible to him in their original composition. Neither the nightingale nor Dyrcona thus perceive the solar beings in their true form. They only see them in the various forms in which they present themselves to each of them, and these forms are merely a product of an imagination curiously attuned to the onlooker: To the nightingale the king and his people first appear as birds (both the nightingale's equal and its enemies) and later as the kind of tree associated with the *topos* of the paradisiacal garden, which offers the perfect setting for the nightingale's lovely song. In introducing themselves to Dyrcona, the king and his subjects change their appearance from a tree and a pomegranate into a dwarf and a multitude of tiny individuals only to undergo yet another metamorphosis into a young man of medium stature (Dyrcona's equal). This final transformation is worth looking at more closely.

The dwarf and his tiny subjects engage in a wild dance and come to represent an enormous giant. Gradually, the individual parts of this giant move more closely together until the outsized figure is reduced to the form of a mid-sized man. Dyrcona comments on this process in the following terms:

but what I found quite marvellous was that the joining of all the component parts of this perfect microcosm took place in the twinkling of an eye. Some of the most agile of our little dancers had leaped up to the height and into the posture essential to the formation of the head; some, warmer and less refined, formed the heart; and some, much heavier, only went into the making of the bones, the flesh, and the fat.[35]

mais ce que je trouvay de bien merveilleux, c'est que la liaison de toutes les parties que achevèrent ce parfait microcosme se fit en un clin d'oeil. Tels d'entre les plus agiles de nos petits danseurs s'élancèrent par une capriole à la hauteur, et dans la posture essentielle à former une teste; tels plus chauds et moins déliez, formèrent le coeur; et tels beaucoup plus pesans, ne fournirent que les os, la chair, et l'embonpoint.[36]

He then goes on to describe the final stage of the metamorphosis:

When this tall, handsome young man was quite completed, ..., I saw the King of all those peoples, of whom he was a hotch-potch [*un cahos*], entering through his mouth. It seems to me, furthermore, that he was inhaled into the body by its own respiration. All this mass of little men had not previously given any sign of life, but as soon as it had swallowed its little King, it no longer felt like anything but

[34] *Soleil*, 145.
[35] *Other Worlds*, 160.
[36] *Soleil*, 141.

> one being. He considered me for some time and having apparently collected himself as he looked at me, he drew near to me[37]

> Quand ce beau grand jeune Homme fut entièrement finy, quoy que sa prompte construction ne m'eust quasi pas laissé de temps pour remarquer aucun intervale dans son progrez, je vis entrer, par la bouche, le Roy de tous les peuples dont il estoit un cahos, encor il me semble qu'il fut attiré dans ce corps par la respiration du corps mesme. Tout cet amas de petits Hommes n'avoit point encor auparavant donné aucune marque de vie; mais si-tost qu'il eut avalé son petit Roy, il ne se sentit plus estre qu'un. Il demeura quelque temps à me considérer; et s'estant comme apprivoisé par ses regards, il s'approcha de moy, me caressa, et me donnant la main ...[38]

The description of this metamorphosis serves two purposes: First of all, it gives an atomistic definition of the human body as a composition of a great number of different parts by presenting Dyrcona's equal and mirror image as a microcosm of tiny organisms. In doing so, it demonstrates the theory of the fundamental equality of matter underlying different phenomena, i.e., the theory which to some extent justifies Cyrano's fiction by maintaining that matter, Gonsales' *excellente comédienne*, in spite of her different costumes, remains the same; that appearances do not matter. But then the metamorphosis also translates the identity of the king and his subjects into the image of the Leviathan: By enacting the incorporation of all subjects into the sovereign's body and presenting the sovereign as both the soul and image of this body politic, it reenacts the famous allegorical picture that forms the frontispiece to Hobbes' work. The solar king and his subjects are imagined in the quintessential image of the European absolutist state.[39] This shows that the representations with which the solar creatures are said to imagine themselves into being in order to become manifest to Dyrcona are the kind of representations created by a distinctly European imagination. They are Dyrcona-Cyrano's own projections. In these acts of self-manifestation, the argument of the limited creativity of human imagination (and, consequently, the limits of Cyrano's fiction when it comes to representing an alternative world) is spelled out in the most unmistakable terms.

The argument receives further support through the distinction that is drawn between light and darker solar regions: For the most part, Dyrcona's solar journey takes place in the darker regions. It is only when he first arrives on the Sun that he finds himself in the *lumineuse campagne* of the light solar regions and completely loses his orientation: He feels as though he were walking on nothing (*soutenus de rien*), and there is nothing his eyes can rest on ("cette terre transparante qu'ils [i.e., mes yeux] pénétroient ne les pouvant soutenir")[40]. He counts his steps carefully and feels relieved when he reaches the darker regions, to which his body seems to be secretly attuned.[41] Immediately, he

---

37 *Other Worlds*, 160. My Addition.

38 *Soleil*, 141-42.

39 The image of the absolute sovereign as a body composed by his subjects forms the frontispiece to the first edition of Hobbes' *Leviathan* of 1651.

40 *Soleil*, 137. Explanation mine.

41 "I felt greatly overjoyed and I imagined that this joy must proceed from a secret sympathy which my body still felt for its opacity." (*Other Worlds*, 155)

falls into a deep sleep. On awaking, he suddenly finds a world with characterizable inhabitants and topographical features. His first encounter is with the king and his nation; as natives of the light regions of the Sun, they are invisible to Dyrcona in their original form just like their place of origin. They only become visible through metamorphosis when they present themselves in surrogate form: The imperceptible world becomes visible by materializing into images of Dyrcona-Cyrano's imagination.

The projections continue in Dyrcona's subsequent encounters with the inhabitants of the Sun. Yet in contrast to the king and his subjects, who originate from the Sun and transform themselves into forms that are perceptible for Dyrcona, the inhabitants of the darker solar regions all originate from the Earth; their forms are therefore already terrestrial.[42] It appears that, parallel to the surrogate appearances of the king and his nation, the entire description of the Sun constitutes some kind of surrogate picture of a world that is imperceptible as well as unimaginable in its otherness. Even the topographical features that characterize this part of the Sun have been imported from the Earth. The forest of talking trees has grown from seedlings of the mythical oaks of Dodona that were carried to the Sun in the stomach of an eagle and vomited out undigested:

> a great eagle, ... was feeding upon the acorns which their branches [i.e., the branches of the oak-trees of the forest of Dodona] offered it. ... Tired of living in a world where it had suffered so much, it took flight to the sun and made such a successful trip that it finally reached this luminous globe, where we are now. But on its arrival, the heat of the climate made it vomit and it brought up a number of still undigested acorns. These acorns germinated and from them there grew the oak trees which were our ancestors.[43]

> une grande Aigle ... se repaissoit du gland que leurs rameux luy fournissoient, quand un jour, ennuyée de vivre dans un Monde où elle souffroit tant, elle prit son vol au Soleil et continua son voyage si heureusement qu'enfin elle aborda le globe lumineux où nous sommes; mais à son arrivée la chaleur du climat la fit vomir; elle se déchargea de force gland non encor digéré; ce gland germa, il en crut des Chesnes qui furent nos ayeuls.[44]

The transposition of images rendered here in narrative terms is integrated into an allegorical tableau that illustrates the nature of the imagination within a mental landscape. After what we said about the techniques of visual figuration in Cyrano's text and their self-referential rhetoric, a discussion of this allegory will offer a suitable conclusion to this chapter: The allegory not only provides a graphic illustration of the interplay between memory and imagination as well as of the dependence of the imagination on sensual impressions, it also reenacts the figuration of Cyrano's fiction itself. It thus presents the metapoetic commentary to this fiction in clear terms.

When Dyrcona is roused from his sleep at the Lake of Sleep, Campanella offers to show him a rare spectacle: Flowing from the lake are five streams, which are named after the five senses. At some point, these five streams flow together only to separate

---

42 At least this is true in those cases where the narrative refers to their first origin.

43 *Other Worlds*, 189. Explanation mine.

44 *Soleil*, 166.

again into three great rivers named after Memory, Imagination and Judgment. Just like the streams of the senses, each river features characteristics that symbolize the particular mental faculty the river stands for. By the river of Memory, for instance, various birds sing their memorized songs by day and drink the thick vapor of the watery region by night;

> but their dyspeptic stomachs find this so hard to digest that in the morning [when they think they have converted it into their own substance,] one sees it all spilling out of their beaks again, in as pure a state as when it was in the river.[45]

> mais leur estomach cacochime la digère si mal qu'au matin, quand ils pensent l'avoir convertie en leur substance, on la voit tomber de leur bec aussi pure qu'elle estoit dans la rivière.[46]

In addition to illustrating the concept of memory in metaphoric terms, the vomiting birds, whose dyspeptic stomachs return their food in its original state, also recall the eagle from the forest of Dodona that spilled the undigested memory of an old legend onto the blank surface of the Sun. There are other such quotations from Dyrcona's narrative in the description of the river of Imagination; its golden fluid carries human fantasies of all kinds:

> The fishes it supports are remoras, sirens, and salamanders. And instead of gravel one finds in its bed those stones that Pliny speaks of which make one heavy when one touches them one side and light when one touches the other. I also noticed there some of those others, from which Gyges had a ring made, that render one invisible. But above all there are a great number of philosopher's stones shining in the sand there. Along the banks were a multitude of fruit trees, particularly those which Mohammed found in paradise. The branches swarmed with phoenixes, and I observed seedlings of that apple tree from which Discord plucked the fruit she threw at the feet of the three goddesses. Cuttings had been grafted on to them from the garden of the Hesperides.[47]

> Le poisson qu'elle nourit, ce sont des Remores, des Syrènes et des Salemandres; on y trouve, au lieu de gravier, de ces cailloux dont parle Pline, avec lesquels on devient pesant quand on les touche par l'envers, et léger quand on se les applique par l'endroit. J'y en remarquay de ces autres encor, dont Gigès avoit un anneau, qui rendent invisibles; mais surtout un grand nombre de pierres Philosophales éclatent parmy son sable. Il y avoit sur les rivages force arbres fruitiers, principalement de ceux que trouva Mahomet en Paradis; les branches fourmilloient de Phénix, et j'y remarquay des sauvageons de ce Fruitier où la Discorde cueillit la pomme qu'elle jetta aux pieds des trois Déesses; on avoit enté dessus des greffes du jardin des Hespérides.[48]

Cyrano's allegory of imagination quotes a variety of old myths and stories for the famous fantasies they contain; it, moreover, quotes Cyrano's own fiction: The remoras

---

45 *Other Worlds*, 219. I have added a translation of the passage omitted by Strachan.
46 *Soleil*, 188.
47 *Other Worlds*, 219.
48 *Soleil*, 188-89.

and salamanders swimming in the river allude to the fight between the remora and the salamander that Dyrcona witnesses on the Sun. And the many phoenixes swarming in the trees of Paradise recall the one Dyrcona converses with after his encounter with the king. There, the bird tells him that after spending one hundred years on Earth, the Phoenix flies to the Sun.

Within the context of these fairly straightforward quotations, other parallelisms emerge where the episodes in Cyrano's fiction that correspond to the images floating in the golden stream have undergone a transformation. But still, may not the magnet that propels Elija's chariot to the Sun be a scientific adaptation of Pliny's magical stones? And when the king and his nation metamorphose into visibility, is this not the inversion of the stones that have the power to make people invisible? Similarly, the fruit trees that Mohammed found in Paradise and the wildings of that tree in the garden of the Hesperides that produced the apple of discord, all of which Dyrcona here sees floating in the river, seem to be reproduced in that other paradisiacal tree with that other 'apple', the pomegranate, hanging on it, which is represented by the king and his subjects. Just as all of these trees appear to be mere quotations of the Biblical tree and its fruit that Dyrcona almost comes to taste of on the Moon, the Tree of Knowledge. Moreover, is not this river "of pure potable gold"[49] with all these images floating in it just an image of the Sun or, more precisely, an image of Cyrano's sun, of his fiction and text? And the dying philosopher lying in the stream with his head inflated from an overload of images, who moves Dyrcona with so much pity, is not he, too, just an image of Dyrcona-Cyrano himself, his head cluttered with fantasies?[50] and thus another image of the explosion that occurred in Cyrano's name and which turned it into its anagram 'Dyrcona'?

The allegorical landscape around the Lake of Sleep thus provides us with a key to the composition of Cyrano's *pot pourry* by offering us a look into Cyrano's mind when this mind is in the process of creating the fiction of the *Soleil*; this allegory represents the poetics of the text. It shows us the sources of Cyrano's imaginative creations as they flow through his mind, shows us how images are remembered, adopted and transformed to be integrated into the flow of Cyrano's own narrative. In so doing, it visualizes the interaction between recollective and innovative productions of the imagination as a continuum:

> Each of these broad rivers divides into an infinite number of channels which merge into one another. I noticed that when a wide stream from Memory approached a smaller one from Imagination it immediately extinguished the latter, but if, on the other hand, the stream from Imagination was bigger, it dried up that of Memory. So as these three rivers, both in their main channels and in their tributaries, always

[49] *Other Worlds*, 219. – "de pur or potable" (*Soleil*, 188)

[50] The episode may allude to the accident that eventually brought about Cyrano's death: As Lachève relates in the biography included in his edition of the *Oeuvres libertines*, Cyrano either knocked his head against a beam or the beam crashed on his head. He never recovered from this accident and in 1654 died from its consequences a little more than a year after it had happened. Though the exact dates of composition of the *Soleil* are not known, what is known is that in 1653 the work was not yet finished and that Cyrano revised it before his death. Accordingly, he may well have included this episode in the face of death. For the account of the accident and the estimated dates of composition of the *Soleil*, see Lachèvre's introduction to *Les oeuvres liberines,* XCI and 1-2.

> run close beside one another, wherever Memory is strong Imagination is reduced and the latter swells as the former abates.[51]

> Chacun de ces deux larges fleuves se divise en une infinité de bras qui s'entrelassent; et j'observay que quand un grand ruisseau de la Mémoire en approchoit un plus petit de l'Imagination, il éteignoit aussi-tost celuy-là; mais qu'au contraire, si le ruisseau de l'Imagination estoit plus vaste, il tarissoit celuy de la Mémoire. Or comme ces trois fleuves, soit dans leur canal, soit dans leurs bras, cheminent toûjours à costé l'un de l'autre, par tout où la Mémoire est forte, l'Imagination diminuë; et celle-cy grossit à mesure que l'autre s'abaisse.[52]

The waters of Memory and Imagination are not clearly distinguishable. Yet whether the fictions of the imagination and of memory lean more to the one or to the other side of the continuum, that is, whether they correspond more to recollective fantasies than to imaginative recollections or vice versa, they are always guided by the senses, which are their only source.

In a note to the text, Frédéric Lachèvre quotes a critical comment which appeared in the first edition of the *Soleil* (in the *Nouvelles Oeuvres* of 1662). It suggests that Cyrano's tableau is a representation of animal spirits (*esprits*) swimming in the cavities of the brain. The function of these spirits is to assume the impression of objects through the nerves (employed in the service of the senses) and to transport the impression to the three faculties of the soul.[53] Indeed, I think that Cyrano's watery landscape contains more than a faint echo of the physiology of the brain offered by Descartes in the *Traité de l'homme*, where these animal spirits (Descartes, too, calls them *esprits*) play a central role; though – in keeping both with the Cartesian theory of memory and the free use Cyrano makes of his sources throughout the *Lune* and the *Soleil* – the memory is a free borrowing.[54] Therefore, rather than recapitulating Descartes' highly puzzling and in many respects incongruous account of the organic functioning of the brain and, more particularly, the mechanics of retention and reconstruction involved in the Cartesian theory of memory, I will pick out those aspects of this theory which Cyrano used for his poetics of the imagination and show in what respects they offer a theoretical description of his intertextual practice in the two novels.

Following the ancient Galenic tradition, Descartes explains all cerebral activity, including memory, dreaming and imagination, by means of animal spirits: Swimming through the cavities of brain and nerves, the spirits impress traces of sense impressions on the internal part of the brain, which is the seat of memory. The traces, which are formed in the brain pores by the spirits wriggling through them, are retained in the

---

51 *Other Worlds*, 220.

52 *Soleil*, 189.

53 *Ibid.*, 188, note 1. In the same note, Lachèvre mentions Ovid and Lucretius as likely sources for Cyrano's allegory.

54 The reference is implicitly given in the text, where Descartes' authority is invoked several times: accordingly, it is Descartes' books that Dyrcona takes with him to his friend's estate before his arrest and subsequent imprisonment (the books are taken for books of black sorcery). Moreover, just before they arrive at the Lake of Sleep, Campanella explains the purpose of his journey to Dyrcona: to meet the great philosopher Descartes, who has recently arrived on the Sun. It is when they finally meet Descartes in person that the *Soleil* comes to an abrupt end.

patterns of the pores. Yet rather than being permanently preserved, the patterns, and with them the memories, keep changing depending on the context in which they are reconstructed. In contrast to theories which considered memories to be a collection of safely stored items, Descartes describes them in terms of reconstructions rather than reproductions. This means that memories never, or only in exceptional circumstances, faithfully correspond to the original experiences they stand for.[55]

When Cyrano adapts memories of classical topoi and well-known stories by recontextualizing them and presenting them from a new perspective, his practice is thus largely in keeping with Descartes' physiological account of *re-membering* earlier experiences. Just as in Cyrano's allegory, the Cartesian brain is an organ bathed in fluids, so that the figures retained there are in constant flux. But the parallelisms do not end here: Since, according to Descartes, memory traces are stored superpositionally and not laid down independently, memories are often blended. The faculty of remembering is therefore closely connected to that of imagination, or rather the other way round: 'imagination ... is the work of memory rather than a separate capacity, ... the same mechanics operates in ordinary imagining and remembering. There are differences only of degree in the confusion of parts in figures which spirits trace." "Traces are distinct only when reconstructed. But it 'usually happens' (*le plus souvent*) that 'several different figures are traced in this same region of the brain almost equally perfectly' ... When this occurs,

> the spirits will acquire a [combined] impression of them all [*quelque chose de l'impression de chacune*], this happening to a greater or lesser degree according to the ways in which parts of the figures fit together. ... It is thus that chimeras and hypogryphs are formed in the imagination of those who daydream, that is to say who let their fancy wander listlessly here and there without external objects diverting it and without the fancy's being directed by reason.[56]

We could thus say that creations of the imagination are caused by deviations in the process of mnemonic reconstruction. It therefore makes sense that, of the three streams in Cyrano's mental landscape, the one attributed to Memory should be the largest by far, since memories are the source of all imaginary creations. Further, Descartes' explanation of the process of dreaming immediately corresponds to that of imagining, so that dreaming, too, is the work of memory. Accordingly, dreams

> differ in nothing from the ideas which, as I said earlier, are formed in the imagination of those who daydream, except perhaps that the images formed while we are sleeping can be much more distinct and lively than those formed when we are awake.

---

55 For this view of memory as a process, see John Sutton's original reappraisal of Descartes' treatise: *Philosophy and Memory Traces: Descartes to Connectionism* (Cambridge: Cambridge University Press, 1998). Sutton points out that according to Descartes' later work *Les passions de l'âme*, wonder is such a special case in which the traces left by the spirits are particularly strong and are therefore preserved in the way in which they were formed. (71)

56 *Ibid.*, 61-62. Additions J.S.

> ils ne different en rien de ces idées que i'ay dit cy-dessus se former quelquefois dans l'imagination de ceux que révent étant éveillez, si ce n'est en ce que les images qui se forment pendant le sommeil, peuvent estre beaucoup plus distinctes & plus vives, que celles qui se forment pendant la vielle.[57]

Descartes' triple juxtaposition of remembering, imagining and dreaming closely parallels the same juxtaposition in Cyrano's narrative. However, two things are more important for Cyrano's metanarrative argument than the parallelism as such: First of all, the Cartesian argument that memories do not remain the same and that their reconstruction is context-dependent, and, secondly, the physiological explanation Descartes offers for the way memories form the source material in imaginative creations. The act of remembering is represented as a creative process, a process of which imagining is then merely a more extreme form.

And yet, to call the processes of remembering and imagining 'creative' is, at least in connection with Descartes' theory, not entirely legitimate since, as Sutton has shown, Descartes explains memory as an exclusively organic process: "[He] does not take memory to be a *mental* capacity at all, for it does not require consciousness or soul. God could (and for us has) joined in puzzling mixture a rational soul to *L'Homme*'s living body-machines, but their life processes operate independently of that union." "The soul may be involved, when it is attached to the machine: but it is not necessary for memory operation."[58] Similarly, the third river in Cyrano's allegory, the river that carries the water of Judgment and which, though smaller than the others, is also "profound", runs its separate course even while it stays close to the intertwining rivers of Memory and Imagination. Again, we may translate this into a definition of the text: To the extent that the *Soleil* merely recounts Dyrcona's dream, it is adequately represented as the purely organic process in which the waters of Memory and Imagination mingle; yet since the *Soleil* is also the artistic creation of its author, Cyrano, we should regard it as a mixture of recollective imaginings and imaginative recollections which are guided and shaped by the sobering influence of Cyrano's 'profound' judgment. The ass is also a noble princess.

---

57 *Oeuvres de Descartes*. Ed. Charles Adam and Paul Tannery (Paris: Vrin, 1967), vol. 11, 198. English translation mine.

58 *Philosophy*, 59 and 74.

5

Metaphors as Systems of Thought: Fontenelle, Cyrano, Wilkins and Huygens

Seventeenth-century philosophers typically understood the concept of plural worlds not just as a plurality of worlds comparable to the Earth in topographical terms, but as a plurality of inhabited worlds, inhabited moreover by forms of rational life. The presence of rational creatures constituted the most important term of comparison between the Earth and other potential worlds. At the beginning of the preceding chapter, I said that the theory of the plurality of worlds is in many respects a direct consequence of Copernican theory; this statement now needs to be more clearly specified. While it is true that with the Earth's recategorization from the center of the cosmos to a planet, the possibility of an essential similarity between the Earth and the other planets suggested itself nearly automatically, this did not automatically turn the possibility into a fact. Nor did it answer the question as to the degree of comparability. And once the Earth's participation in the planetary dance around the Sun was established, it was still quite a different matter to turn from a plurality of worlds in our solar system to a universe of innumerable worlds by identifying the fixed stars as other suns and claiming that each of these suns had its own – invisible – planetary system orbiting around it. The cosmological theory developed by Copernicus was clearly the first step toward such a cosmology, but it was not much more.

Proponents of the theory of the plurality of worlds therefore often founded their theory on a couple of *a priori* metaphysical principles of religious or philosophical origin. One such principle was the concept of the uniformity of nature, which holds that there is nothing unique in the universe, but that everything belongs to a generic type. This concept is explicitly formulated in Lucretius' *De natura rerum*, a work that was extremely influential in reintroducing the idea of plural worlds in the seventeenth century.[1]

Another principle that was often used in this connection is the principle of plenitude. Based on the hypothesis that the creative power is infinite, it presupposes the full realization of conceptual possibility into actuality.[2] Although, in its original Platonic formulations, and even more so in its Christian reformulations, the principle of pleni-

---

1 Stephen Dick, *Plurality*, 11.

2 Arthur Lovejoy, *Chain of Being*, *passim*.

tude represented a strictly vertical order that clearly adhered to the geocentric cosmos and the theological hierarchy this cosmos was believed to represent,[3] another tradition also existed. Lucretius, for instance, assigns creative power to Nature rather than to God and turns the concept of 'fullness' into that of fecundity. He argues that "when abundant matter is to hand, and no thing and no cause hinders, things must assuredly be done and completed."[4] In this way, the strictly vertical order of the Platonic-Christian principle becomes a multidimensional one.

Of clearly religious origin are two other arguments that were regularly used to support the theory of the plurality of worlds: The first is the teleological principle which holds that nothing has been created without purpose and that, therefore, the other planets around the Sun cannot be wastelands, but must offer habitation to different forms of life. Moreover, so the argument typically runs, since all those many stars that remain invisible to the unaided eye cannot have been created for us, they must be a source of life in their own right by shining on other inhabited worlds. Related to this argument is the Christian concept that God's Creation finds its ultimate fulfillment in the contemplation of rational beings, i.e., God's creatures. In view of the fact that great parts of the universe are beyond our visual reach, this concept necessitates the existence of rational beings in remote parts of the universe, whose visual reach covers what is invisible from the Earth.[5] Finally, the teleological principle is related to that of economy, which holds that Nature achieves as much as possible with the least possible means, that, given the opportunity, it realizes different ends by a single means. Therefore, though it is good that the stars give human beings on Earth pleasure and that they offer them points of reference to orient themselves in space and time, it is better if they also serve as suns to their own planets.[6]

---

3 See p. 106, note 1, *supra.*

4 Quoted in Dick, *Plurality*, 11.

5 In the *Cosmotheoros* of 1698, Christiaan Huygens repeatedly turns to this argument. In reference to those who claim that the universe has been created exclusively for the inhabitants of the Earth, he writes: "And these Men themselves can't but know in what sense it is that all things are said to be made for the use of Man, not certainly for us to stare or peep through a Telescope at; for that's little better than nonsense. Since then the greatest part of God's Creation, that innumerable multitude of Stars, is plac'd out of the reach of any man's Eye; and many of them, it's likely, of the best Glasses, so that they don't seem to belong to us." Huygens therefore suggests that the multitude of stars invisible to us could be seen by the inhabitants of other planets.

Later, after describing the diverse plants and animals on other planets, he points out the need of "placing some Spectators" on them as well. "For all this Furniture and Beauty the Planets are stock'd with seem to have been made in vain, without any design or end, unless there were some in them that might at the same time enjoy the Fruits, and adore the wise Creator in them." Cf. *The Celestial Worlds Discovered* [*Cosmotheoros*]. Facsimile reprint of edition by Timothy Childe of 1698 (London: Frank Cass, 1968), 7-8 and 36-37.

6 Its economy of means was often claimed to be one of the major advantages that the Copernican system had over the old geocentric universe. Accordingly, Huygens says: "This is the now commonly receiv'd System, invented by *Copernicus*, and very agreeable to the frugal Simplicity Nature shows in all her Works." He also sharply criticizes Athanasius Kircher for squandering so much material on the 'furniture' of the planets without allowing them to be inhabited. (*Ibid.*, 13 and 101-5)

In the *Entretiens sur la pluralité des mondes*, the work that served Huygens as his principal source, Fontenelle also praises the beautiful simplicity of the Copernican system. After telling the

To conclude this list, there is the principle that is most fundamental of all, since it establishes the validity of all the other principles: It posits the essential rationality of the order of the universe and the congruence of that rationality with the interpreting mind.[7] Only on this premise can any hypothesis about the universe be formulated at all. The arbitrariness of these and similar principles used in connection with the hypothesis of the plurality of worlds is evident, especially since some of them contradict each other. Yet their usefulness lies in their offering a framework that allows philosophers to formulate this hypothesis in the first place, and to formulate it convincingly. It is the notion of the framework, of the structure for thought, which is going to occupy us here. I want to show that certain metaphors used to illustrate the concept of plural worlds offer similar frameworks for the imagination.

Once the theory of the plurality of worlds has been established in principle, the question of how we are to imagine these worlds arises. Huygens and Fontenelle approach this question from opposite sides; yet both approaches rest on the kind of *a priori* principles presented above. In his *Cosmotheoros*, Huygens bases his argument on the principle of uniformity in nature and argues that even while Nature may have made things on other planets infinitely different from those on Earth and "such as neither our Understanding nor Imagination can conceive", it does not follow that because this variety "may be Infinite, and out of our comprehension and reach, that therefore things in reality are so."[8] The principle of uniformity in Nature is here paradoxically deduced from the principle of Nature's creative freedom. Huygens uses this as a licence for imagining the worlds on other planets with their plants, animals and

---

Marquise about the confusing complexities of the Ptolemaic cosmological system, Fontenelle turns to that devised by Copernicus: "'I'm now going to propose a different one to you which satisfies all ... for it's one of a charming simplicity, which alone would make it preferable.' 'It would seem,' the Marquise interrupted, 'that your philosophy is a kind of auction, where those who offer to do these things at the least expense triumph over the others.' 'It's true,' I replied, 'and it's only by that means that one can catch the plan on which Nature has made her works. She's extraordinarily frugal. Anything that she can do in a way which will cost a little less, even the least bit less, be sure she'll only do it that way. This frugality, nevertheless, is quite in accord with an astonishing magnificence which shines in all she does. The magnificence is in the design, and the frugality in the execution.'" Cf. Bernard le Bovier de Fontenelle, *Conversations on the Plurality of Worlds*, ed. Nina Rattner Gelbart, trans. H. A. Hargreaves (Berkeley: University of California Press, 1990), 15. Unless otherwise noted, references in English refer to this edition, henceforth cited as *Conversations*.

7 E. D. James, "Fontenelle's *Entretiens sur la pluralité des mondes* and Their Intellectual Context," in Charles G. S. Williams, ed., *Actes de Columbus: Racine. Fontenelle, Entretiens sur la pluralité des mondes: Histoire et littérature* (Paris: Papers on Seventeenth-Century French Literature, 1990), 137. See also the seminal study by Amos Funkenstein, *Theology and the Scientific Imagination from the Late Middle Ages to the XVIIth Century*, (Princeton: Princeton University Press, 1986), where he discusses the relationship between the natural sciences and three central attributes of the divine (omnipotence, omnipresence and providence) and the development of this relationship from the late Middle Ages to the seventeenth century. Of particular interest in our context is his third study, which focuses on God's omnipotence and the concept of natural laws.

8 *Cosmotheoros*, 22. – "Neque, quia posset essere immensa, resque illae ab intellectu, & comprehensione nostra penitus remotae, idcirco necesse esse, ut reipsa tales existant." *Christiani Hugenii Kosmotheoros, sive De Terris Coelestibus, earumque ornatu, conjecturae*, in *Oeuvres complètes de Christiaan Huygens* (The Hague: Martinus Nijhoff, 1944), vol. 21, 703.

rational inhabitants in forms almost identical to those on Earth, some minor variations in Nature's handiwork granted. He justifies this hypothesis with the argument that the way things are on Earth is so perfect that there is little room for variation, since Nature cannot have created the other worlds as anything less than perfect.[9]

In the *Entretiens sur la pluralité des mondes* (*Conversations*), Fontenelle departs from the contrary, more challenging premise: He bases his hypothesis of the nature of plural worlds and their inhabitants on the principle of the "infinite diversity that Nature placed in her works", arguing in terms of what we could call the Lucretian version of the principle of plenitude.[10] The problem this infinite diversity poses for the imagination represents one of the principal themes of the *Conversations*. It is clearly stated by the Marquise in the conversation on the third evening: "My reason is pretty well convinced, ... but my imagination's overwhelmed by the infinite multitude of inhabitants on all these planets, and perplexed by the diversity one must establish among them".[11] The narrator confirms this and offers his solution to the problem by means of an analogy:

> It is not up to the imagination to attempt to picture all that. ... It is not proper for the imagination to go any farther than the eyes can. One may only perceive by a kind of universal vision [*vûë universelle*] the diversity which Nature must have

---

9 The following argument about the outward appearance of rational planetarians offers a good example: "The Stature and Shape of Men here [i.e., on Earth] does show forth the Divine Providence so much in its being so fitly adapted to its design'd Uses, that it is not without reason that all the Philosophers have taken notice of it nor without probability that the Planetarians have their Eyes and Countenance upright, like us, for the more convenient and easy Contemplation and Observation of the Stars. And the Wisdom of the Creator is so observable, so praiseworthy in the position of the other Members; in the convenient situation of the Eyes, as Watches in the higher Region of the Body; in the removing of the more uncomly parts out of sight as 'twere; that we cannot but think he has almost observed the same Method in the Bodies of those remote Inhabitants. Nor does it follow from hence that they must be of the same shape with us. For there is such an infinite possible variety of Figures to be imagined, that both the Oeconomy of their whole Bodies, and every part of them, may be quite distinct and different from ours" (73-74, explanation mine). Though with this final statement Huygens grants a little room for variation, it is clearly not much.

Huygens' argument here is strikingly analogous to Leibniz's elaboration of the coincidence of the real world with the best of all possible worlds in the high priest Théodore's visionary dream in the *Théodicée*. This coincidence corresponds to the metaphysical necessity which requires that "if a rational God created the world, its order had to agree with and be determined by a rational logic." The world created by Jupiter could not but be the best of all worlds possible. Cf. Gottfried Wilhelm Leibniz, *Essais de Théodicée, sur la bonté de dieu, la liberté de l'homme, et l'origine du mal*, ed. J. Brunschwig (Paris, 1969), 361-62. See also Funkenstein's discussion of divine omnipotence and natural laws, where he confronts Leibniz's position with those of Descartes and Spinoza, in *Theology*, part III, as well as Timothy Reiss's discussion of this passage in *The Meaning of Literature* (Ithaca and London: Cornell University Press, 1992), 171-73.

10 *Conversations*, 6.

11 "Ma raison est assés bien convaincuë, dit la Marquise[,] mais mon imagination est accablée de la multitude infinie des Habitans de toutes ces Planetes, et embarrassée de la diversité qu'il faut établir entre eux". Cf. *Entretiens sur la pluralité des mondes*, ed. Alexandre Calame (Paris: Librairie Marcel Didier, 1966), 95. Addition mine. All quotations in French are taken from this edition, henceforth cited as *Entretiens*.

placed among all the worlds. All faces in general are made on the same model, but those of two large societies – European, if you like, and African – seem to have been made on two specific models, and one could go on to find the model for each family. What secret must Nature have possessed to vary in so many ways so simple a thing as a face? In the universe we're no more than one little family whose faces resemble one another; on another planet is another family whose faces have another cast.

We can suppose the differences increase according to the distance one travels.[12]

Ce n'est pas à l'imagination à prétendre se le representer, répondis-je, elle ne peut aller plus loin que les yeux. On peut seulement appercevoir d'une certaine vûë universelle la diversité que la Nature doit avoir mise entre tous ces Mondes. Tous les visages sont en general sur un même modele; mais ceux de deux grandes Nations, comme des Européens, si vous voulés, et des Afriquains ou des Tartares, paroissent être faits sur deux modeles particuliers, et il faudroit encore trouver le modele de visages de chaque Famille. Quel secret doit avoir eu la Nature pour varier en tant de manieres une chose aussi simple qu'un visage? Nous ne sommes dans l'Univers que comme une petite Famille, dont tous les visages se ressemblent; dans une autre Planete, c'est une autre Famille, dont les viages ont un autre Air.

Apparemment les differences augmentent à mesure que l'on s'éloigne[13]

Although Fontenelle's model for nature's diversity is based on deductive and inductive reasoning, its aim is to aid imagination.[14] By presupposing some resemblance between the inhabitants of other worlds and human beings on Earth, Fontenelle establishes a fundamental comparability between them. Nature's diversity is therefore not total. The 'universal vision' that the narrator advocates illustrates this diversity in an analogy to the diversity of facial features among human beings and the order in which the differences among them are arranged. The analogy is substantiated by the argument that these differences increase with increasing distance and by the extension of this pattern into space through induction. Accordingly, Fontenelle assumes the pattern of differences to be governed by yet another rule that controls the principle of the infinite diversity and forces it into conceivable bounds.

The analogy thus offers both an image and a pattern of relations. Yet while the image is useful for its metaphoric value in as much as it stands in place of the images the imagination cannot produce, the pattern of relations is more important. Fontenelle here offers the Marquise (and, by implication, his readers) a framework for the hypothetical system of the universe he is proposing, a structure in which to think his theory. The image is only important to the extent that it makes this theory imaginable through metaphor. It is the pattern of relations connecting the different parts of the image which makes the diversity conceivable in principle. Rather than reducing alterity to the same, as Huygens does, Fontenelle tries to capture it *as* alterity, but in a system in which it is accessible to reason, in a system that, moreover, might remove any threat inherent in the idea of uncontrollable alterity. As Jean Dagen has aptly put

---

12 *Conversations*, 45-46. Addition mine.

13 *Entretiens*, 95-96.

14 Robin Howells, "The Principle of Mobility in Fontenelle's *Entretiens sur la pluralité des mondes*," in *French Studies* 46, no. 2 (1992): 134.

it: “En somme, ce qui intéresse Fontenelle et dont il fait son livre, ce n’est pas véritablement le monde, mais la pensée du monde. Comment peut-on concevoir un monde. Voilà qui importe plus que l’image à laquelle il est actuellement possible de s’arrêter.”[15]

In the following, I would like to discuss the function of certain metaphors and analogies that keep reappearing in the *Conversations* as well as in some other texts in which the theory of the plurality of worlds is promoted. We will see that they do not just serve to translate the hypothetical cosmology into familiar images (though, to the reader’s delight,[16] they do that too); more importantly, they offer patterns of relations that lend the theory both authority and narrative coherence. In some cases, they also provide what Fontenelle calls ‘universal visions’ that help to make conceivable what is beyond imagination. Accordingly, they have both a representational and a persuasive function. Although I will take examples from a variety of texts, the main focus will be on Fontenelle’s *Conversations* for the way they reflect on their own techniques of representation.

In the narrator’s conversations with the Marquise, the argument repeatedly turns to the inability of the imagination to posit otherness. The most straightforward references of this kind refer to the impossibility of imagining the inhabitants of other worlds in forms radically different from human beings on Earth. Yet the *Conversations* also focus more generally on the imaginative projections that characterize their narrative; such projections are indeed said to be the common basis of representation in general. Thus, the narrator refers to “a certain young girl that has been spotted on the Moon through the telescope some forty years ago who has aged considerably,” and he adds:

> since everyone projects onto an object those ideas with which he is filled. Our Astronomers see on the Moon the faces of young girls, and it is quite possible that if women observed [the Moon], they’d see the handsome faces of men there. As for myself, Madame, I think that I would see you there.

> une certaine Demoiselle que l’on a vûe dans la Lune avec des Lunettes, il y a peut-être quarante ans, que ne soit considerablement vieillie. ...
>
> ... car chacun transporte sur les objets les idées dont il est rempli. Nos Astronomes voyent sur la Lune des visages de Demoiselles, il pourroit être que des Femmes qui observeroient, y verroient de beaux visages d’Hommes. Moi, Madame, je ne sçai si je ne vous y verrois point.[17]

Georges Van Den Abbeele has pointed out that Fontenelle’s concern in the *Conversations* are “the very conditions of representability, not only as a conceptual practice but

---

[15] “Réflexions sur les mondes de Fontenelle,” in *Littératures classiques* 22 (1994): 133.

[16] For Fontenelles’ Horatian objective both to delight and to instruct in his book, see his Preface.

[17] *Entretiens*, 167 and 169. Explanation mine. I have given my own translation of this passage since the English edition by Rattner Gelbart, which I am using elsewhere, only includes the original five conversations of 1686 and not the sixth conversation added later, from which this passage is taken.

For an interesting argument about “figuration as defiguration” in the *Entretiens*, especially the figuration of the feminine face and its role as “figure of figuration itself”, see Georges Van Den Abbeele, “*Fabula est Mundus*: On the Plurality of Fontenelle’s ‘Worlds’,” in Williams, ed., *Actes*, 165-83.

also as a communicative one."[18] Therefore, even while its inadequacy for accurate representation is a constant theme, the imagination plays an important role in this text, both for the narrator and his audience.[19] This double role is illustrated by the narrator in an extended metaphor.

Nature is compared to an opera house: "From where you are at the opera you don't see the stages exactly as they are; they're arranged to give the most pleasing effect from a distance, and the wheels and counter-weights that make everything move are hidden out of sight. You don't worry, either, about how they work. Only some engineer in the pit, perhaps, may be struck by some extraordinary effect and be determined to find out for himself how it was done. That engineer is like the philosopher. ... Whoever sees nature as it truly is simply sees the backstage area of the theater".[20] The philosopher, like the engineer in the pit, seeks to find the physical laws responsible for the spectacle he is witnessing. So far, the narrator's comparison is fairly straightforward. Only when we read it in connection with the following passage in the Preface does it develop its full significance.

The narrator first explains that to follow the argument presented in his text "it's merely required that one read and at the same time *form a clear idea* of what one is reading".[21] Thus, the narrator refers here to the importance of the readers' picturing in their mind the argument he presents. Of the ladies, who, as his female interlocutor shows, are his special audience, all that is required is "the same amount of concentration that must be given to *The Princess of Cleves* in order to follow the plot closely and understand all its beauty".[22] Then the narrator adds:

> Since I had no intention of creating a make-believe system, without any foundation, I've employed verifiable physical tenets, as many as were necessary. But fortunately it happens that on this subject the ideas of physics are pleasing in themselves [*riantes d'elles-mêmes*] and, at the same time that they're satisfying the mind, they provide a *spectacle for the imagination which pleases it* as much as if they had been made expressly for that purpose.
>
> When I found some tidbits that were not entirely in this vein, *I gave them exotic trappings* [*ornemens étrangers*].[23]

---

[18] *Ibid.*, 168.

[19] E. D. James notes the frequency with which the words '*imaginer*' and '*imagination*' occur in the *Conversations*. ("Intellectual Context," 141)

[20] *Conversations*, 11-12. – "Du lieu où vous êtes à l'Opera, vous ne voyés pas le Théatre tout-à-fait comme il est; on a disposé les Décorations et les Machines, pour faire de loin un effet agréable, et on cache à votre vûë ces rouës et ces contrepoids qui font tous les mouvemens. Aussi ne vous embarassés-vous guere de deviner comment tout cela joue. Il n'y a peut-être que quelque Machiniste caché dans le Parterre, qui s'inquiéte d'un Vol qui lui aura paru extraordinaire, et qui veut absolument démêler comment ce Vol a été exécuté. Vous voyés bien que ce Machiniste-là est assés fait comme les Philosophes. ... qui verroit la Nature telle qu'elle est, ne verroit que le derriere du Théatre de l'Opera." (*Entretiens*, 17-20)

[21] *Conversations*, 4. Emphasis mine. – "c'est seulement ne point lire sans se représenter nettement ce qu'on lit." (*Entretiens*, 6)

[22] *Conversations*, 4-5. – "Je ne demande aux Dames pour tout ce Sistême de Philosophie, que la même application qu'il faut donner à la Princesse de Cléves, si on veut en suivre bien l'intrigue, et en connoître toute la beauté." (*Entretiens*, 6)

[23] *Conversations*, 5. Emphasis mine.

> Comme je n'ai pas prétendu faire un Sistême en l'air, et qui n'eût aucun fondement, j'ai employé de vrais raisonnemens de Physique, et j'en ai employé autant qu'il a été nécessaire. Mais il se trouve heureusement dans ce sujet que les idées de Physique y sont riantes d'elles-mêmes, et que dans le même tems qu'elles contentent la raison, elles donnent à l'imagination un spectacle qui lui plaît autant que s'il étoit fait exprès pour elle.
>
> Quand j'ai trouvé quelques morceaux qui n'etoient pas tout à fait de cette espece, je leur ai donné des ornemens étrangers.[24]

Not only does Fontenelle use here again the metaphor of the theatrical stage, he also reverses it: The physical principles that in his conversation with the Marquise are referred to as the reality behind the fiction are here said to possess spectacular qualities themselves and to provide a pleasing spectacle for the imagination. Where they lack such qualities, the narrator acknowledges having fashioned them with 'trappings' of his own imagination. It is in this guise that the physical principles enact a plot comparable to that of a popular novel, full of scandalous revelations of courtly intrigues.

In metaphorical terms, the function of imagination is here defined as that of producing metaphors ('trappings') to illustrate the working principles of the universe as a plot of very human dimensions. The kind of 'trappings' the narrator uses to represent the system of the universe to the Marquise confirms the aptness of his prefatory statement. Accordingly, the metaphors and analogies we will be focusing on in this chapter are all socially or politically motivated. They revolve around two basic systems: On the one hand, there is the political metaphor that includes references to the interaction between individual states as well as references to a monarchy and its hierarchically organized society; on the other hand, there is the biological analogy to the human body composed of innumerable smaller bodies, and its metaphorical representation as innumerable worlds within worlds. At some point, these two systems are combined in one image, a kind of extended Leviathan.

Fontenelle ties the theory of the plurality of worlds to the Cartesian cosmology of vortices, which holds that planets are carried around their suns in a vortex of celestial matter. He describes the vortex around our Sun as a system of neighboring states living peacefully side by side. This comfortable situation was, however, established only after a period of strife and military conquests at the beginning of creation; at that time, larger planets subjugated smaller ones and forced them to turn around them, thus forming a mini-vortex within the larger vortex around the Sun.

> That's how at the world's beginning we made the Moon follow us, because she found herself within the limits of our vortex and totally at our disposal. Jupiter ... was more fortunate and more powerful than we were; there were four little planets in its neighborhood and it subjugated all four.[25]

> C'est ainsi qu'au commencement du Monde nous nous fîmes suivre par la Lune, parce qu'elle se trouva dans l'étendue de notre Tourbillon, et tout-à-fait à notre

[24] *Entretiens*, 6-7.
[25] *Conversations*, 53.

> bienséance. Jupiter ... fut plus heureux ou plus puissant que nous. Il y avoit dans son voisinage quatre petites Planetes, il se les assujetit toutes quattre[26]

When the Marquise fears that Jupiter might draw the Earth into its power, the narrator reassures her that "the planets stay where they are and new conquests are denied them, as they once were to the kings of China".[27] The peaceful cohabitation of larger and smaller planets is considered analogous to the way liquids of different weights settle at different levels. The Marquise in turn translates this chemical analogy into a social contract assigning each member the specific place in the social hierarchy that by its nature suits it best:

> I understand ... that the weights of things regulate the ranks very well. Would to God that there were something similar to regulate us, which would fix people in those positions to which they're suited by nature.[28]

> Je conçois ... que ces pesanteurs-là réglent fort bien les rangs. Plût à Dieu qu'il eût quelque chose de pareil qui les réglât parmi nous, et qui fixât les gens dans les places qui leur sont naturellement convenables![29]

According to the Marquise, the social stratification of people would be just as natural as the gravitational one of liquids; among people on Earth, unfortunately, it remains her utopian dream. The Marquise then projects this order onto the system of the universe when she expresses herself in favor of that cosmology which most respects social hierarchy this condition is best met by the Copernican system. Because of her regard for social boundaries, the Marquise considers the Tychonic system as "affected", since it "exempt[s] the Earth from turning about the Sun when one could exempt no other large bodies." She simply deems it "not so fitting for the Sun to turn about the Earth when all the planets turned around it."[30] Similarly, she would like "the inhabitants of the four moons of Jupiter to be like its colonies, that they'd receive from it their laws and their customs and that consequently they'd give it some sort of homage and regard the great planet with nothing but respect."[31]

This natural order is at times disturbed, however. In some instances the planets behave little better than quarreling children: Reflecting the self-important way in

---

[26] *Entretiens*, 116.

[27] *Conversations*, 54. – "les Planetes se tiennent où elles sont, et les nouvelles conquêtes leur sont défendues, comme elles l'étoient autrefois aux Rois de la Chine." (*Entretiens*, 117)

[28] *Conversations*, 54.

[29] *Entretiens*, 118-19.

[30] *Conversations*, 22. Adaptation mine. – "qu'il y avoit trop d'affectation à exempter la Terre de tourner autour du Soleil, puisqu'on n'en pouvoit pas exempter tant d'autres grands Corps; que le Soleil n'étoit plus si propre à tourner autour de la Terre, depuis que toutes les Planetes tournoient autour de lui". (*Entretiens*, 44)

The notion of appropriateness is related to the principle of economy in nature; a similar argument can already be found in the first book of *De revolutionibus*.

[31] *Conversations*, 56. – "Je voudrois donc, replit-elle, que les Habitans des quatre Lunes de Jupiter, fussent comme des Colonies de Jupiter; qu'elles eussent reçu de lui, s'il étoit possible, leurs Loix et leurs Coûtumes; que par conséquent elles lui rendissent quelque sorte d'hommage, et ne regardassent la grande Planete qu'avec respect." (*Entretiens*, 120-21)

which human beings attribute all motion to the heavens above them, the Moon accuses the Earth of causing the slight imbalances which are really due to her own librations. This the Marquise compares to the way we blame "the others for what we ourselves do. The Earth says: 'It's not I who turn, it's the Sun.' The Moon says: 'It's not I who sway, it's the Earth.' There are lots of errors everywhere.'"[32] Order is preserved with a rigid system of control which guarantees that impostors get the punishment they deserve. The biggest impostor of them all is the Earth: She abandoned her assigned place and usurped the Sun's position at the center of the vortex, forcing the Sun and the other planets to circle round her. Yet divine vengeance interceded, personified by Copernicus, who with God-like power set things right and reestablished the original heliocentric system.

> Copernicus ... lays violent hands on the different circles and solid spheres which were imagined by Antiquity. He destroys the first and breaks the others in pieces. Seized by a noble astronomical fury, he plucks up the Earth and sends her far from the center of the universe, where she was placed, and puts the Sun in the center, to whom the honor rightly belongs. The planets no longer turn around the Earth and enclose her in the circles they describe. If they light us, it's more or less by chance as we meet them in their paths. Everything turns around the Sun now, including the Earth, and as punishment for the long rest she was given, Copernicus charges her as much as he can with the same movements she had attributed to the planets and heavens. At last the only thing left of all this celestial train which used to accompany and surround this little Earth is the Moon that turns around her still.[33]

> Copernic ... fait main-basse sur tous ces Cercles différens, et sur tous ces Cieux solides qui avoient été imaginés par l'Antiquité. Il détruit les uns, il met les autres en pièces. Saisi d'une noble fureur d'Astronome, il prend la Terre et l'envoye bien loin du centre de l'Univers, où elle s'étoit placée, et dans ce centre, il y met le Soleil, à qui cet honneur étoit bien mieux dû. Les Planetes ne tournent plus autour de la Terre, et ne l'enferment plus au milieu du Cercle qu'elles décrivent. Si elles nous éclairent, c'est en quelque sort par hazard, et parce qu'elles nous rencontrent en leur chemin. Tout tourne présentement autour du Soleil, la Terre y tourne elle-même, et pour la punir du long repos qu'elle s'étoit attribué, Copernic la charge le plus qu'il peut de tous les mouvemens qu'elle donnoit aux Planetes et aux Cieux. Enfin de tout cet équipage céleste dont cette petite Terre se faisoit accompagner et environner, il ne lui est demeuré que la Lune qui tourne encore autour d'elle.[34]

In this powerful allegorical scene Fontenelle turns the history of cosmology into a history and genealogy of our solar system that evokes the narrative of our first parents' transgression and subsequent fall from grace. In analogy to Adam and Eve, whose hunger for divine omniscience made them transgress against the place assigned to them, the Earth here exemplifies the pride and blindness of humankind in envisioning itself at the center of creation. By drawing from the narrative of the Genesis,

---

[32] *Conversations*, 26. – "comme nous, qui rejettons toujours sur les autres ce qui est en nous-mêmes. La Terre dit; *Ce n'est pas moi qui tourne, c'est le Soleil.* La Lune dit, *Ce n'est pas moi qui tremble, c'est la Terre.* Il y a bien de l'erreur partout." (*Entretiens*, 52)

[33] *Conversations*, 15-16.

[34] *Entretiens*, 28-29.

Fontenelle cleverly fends off the charges of blasphemy typically brought up against Copernicans for refashioning the system of the universe advocated by the Church: In his account of the restoration of Creation, Fontenelle not only fills in the lacunae of the original narrative, which contains little cosmological information, he also fashions Copernicus as the protector of God's universe and executor of divine law.

On the secular level, the reinstatement of the Sun as ruler and the restoration of political stability take authority from the Sun King (Fontenelle's king) and the concept of the divine right of kings. In exchange, the Sun's rightful rule over the universe in turn confirms the rightfulness of the French monarchy. In this way, Fontenelle's rhetoric dissociates Copernican theory from any revolutionary movement and presents it instead as loyal to both king and Church, thereby soliciting these two powers for their support. Copernicus' powerful intervention conveys the idea of the persuasive power of his theory, and, more urgently, the sense of the driving force of truth coming to light and attaining its right.[35]

Fontenelle was neither the first nor the last to refer to the solar system in the metaphorical terms of a social system. Almost fifty years before him, John Wilkins used this metaphor in his treatise *The Discovery of a World in the Moone*. There Wilkins argues that the Moon is very similar to the Earth in topography and that it is likely to contain a world with inhabitants just like the Earth. He uses Copernican theory to support his claim. In order to disprove the Aristotelian concept of an immutable heaven, Wilkins refers to Tycho Brahe's evidence, which proves that comets are produced beyond the lunar sphere. He then summarizes the view of the philosophers who, in order to salvage the concept of celestial immutability, counter Brahe's argument by claiming that comets are caused only by minor irregularities in the heavens, so minor, that is, as to be negligible. Wilkins uses the metaphor of the cosmos as social system to refute this view:

> But, say some, this doth not argue any naturall alteration in those purer bodies, since 'tis probable that the concourse of many *little vagabond starres* by the union of their beames may cause so great a light. ... For, say they, when there happens to be a concourse of some few starres, then doe many other flie unto them from all the parts of heaven like so many *Bees unto their King*. But 1. 'tis not likely that amongst those which wee count the fixed starres there should be any such uncertaine motions, that they can wander from all parts of the heavens, *as if Nature had neglected them, or forgot to appoint them a determinate course*. 2. If there be such a conflux of these, as of Bees to their King, then what reason is there that they doe not still tarry with it, that so the Comet may not be dissolved?[36]

In presenting and criticizing this theory, Wilkins uses two contrastive social images: first one of instability, then one of stability. By referring to the many little stars said to form a comet as 'vagabonds', he represents them as outlaws for whom there is no place in society. Although such social outcasts exist on Earth, they are inadmissible in the larger cosmos, since their existence would imply that Nature's plan shows serious

---

[35] For a discussion of metaphorical references to truth and the power of its revelation, see Blumenberg, *Paradigmen*, chapter 1.

[36] John Wilkins, *The Discovery of a World in the Moone* (1638) (Hildesheim: Georg Olms Verlag, 1981), 175-76. Emphasis mine.

flaws. Accordingly, Wilkins here argues in terms of the very concept he is so intent on disproving in his larger argument: the dichotomy between the terrestrial and the celestial sphere and the notion of a perfectly ordered heaven.

The second image Wilkins evokes in his refutation involves the paragon of organized societies, the state of bees.[37] Here Wilkins' argument takes exactly the opposite tack as before: While the comparison of the clustering stars with a state of bees would meet the standards of celestial beauty and stability, it is not acceptable for the fact that comets are only transitory phenomena. This time it is the comets, by not corresponding to the behavioral patterns of bees, that do not meet the standard of order set for them. The fact that the process the theory describes does not match the metaphor used to represent it in this case provides the justification for the rejection of the theory.

The second theory on the generation of comets that Wilkins goes about to refute is that which holds that comets

> proceed from the body of the Sun, and that that Planet [i.e., the Sun] onely is [quotation in Latin][38]: *The shop or forge of Comets* from whence they were sent, *like so many spies*, that they might in some short space returne againe, but this cannot be, since if so much matter had proceeded from him alone, it would have made a sensible diminution in his body.[39]

The solar exhalations that, according to this theory, are responsible for the formation of comets are here compared to political spies sent out by the Sun (which, as a result of this process, is reduced to being the 'shop or forge of comets'), so as to evoke the picture of a state eaten up by anxiety and distrust. The criticism made on the literal level, i.e., that this process would cause a sensible diminution of the solar body, is not very convincing if we consider that the comet is only visible by night, in the Sun's absence. It becomes ludicrous once Wilkins advances his own theory about the nature of comets and claims that they are exhalations of the planets! If the production of comets did indeed visibly diminish the solar body, what would happen to the body of a planet, especially one of the smaller ones in our vicinity, if comets were produced through its exhalations? Wilkins' refutation of this theory thus works exclusively through the metaphorical image: A state permeated by corruption and wrong dealings is an unacceptable representation of the cosmos. Along with the refutation of the first theory, it rests on the proposition that the cosmos is an 'incorruptible' system governed

---

37 Bees stirred the interest of quite a number of philosophers, especially in the first half of the eighteenth century, because the organization of their community was considered to offer a paragon of the perfect state; see Bernard Mandeville's *Fable of the Bees* and J. B. Simon's *Le Gouvernement admirable; ou la République des abeilles*. Other studies on beehivorial patterns were written by Maraldi, Swammerdam and Réaumur. Cf. Alexandre Calame's note in *Entretiens*, 100, note 1.

Earlier in the century, observations on bees were made by Maraldi. Moreover, almost an entire volume of Réaumur's *Mémoires* is taken up by a treatise on bees.

38 The comparison is originally by Tycho Brahe, who equally refutes the theory in favor of another; cf. his *Astronomiae instauratae progymnasmata*.

39 *Discovery*, 178. Explanations and emphasis mine. The first edition of Wilkins's treatise, which I am using here in facsimile, mistakenly numbers four consecutive pages with "178" and "179"; this passage is taken from the first page 178.

by law and order which allows the philosopher to take it for granted that certain things just do not happen there. That Wilkins should offer an argument so permeated by metaphor is particularly ironic since, beside being a popularizer of Copernican theory, he was also the author of an *Essay Toward a Real Character and a Philosophical Language*, in which he urged that metaphor be excluded from natural philosophy. Instead, he "sought a new standard of plain, unadorned language for science."[40]

Fontenelle and Wilkins build their theories on the authority and narrative coherence of established religious and political discourses of power that they project onto the system of the cosmos. By using such discourses metaphorically, they exploit the system of relations delineated in these discourses to lend their cosmological argument the desired narrative coherence. Accordingly, metaphor here links scientific discourse with the values and ideological interests associated with specific cultural institutions so as to make its argument convincing.[41] Let me illustrate this transfer of narrative coherence with a further example.

The most remarkable elaboration of the metaphor of a social hierarchy that I have come across appears in an English translation of the *Cosmotheoros*, where the metaphor is used in support of the theory of the plurality of worlds. The translation was published in the same year as the original publication of the *Cosmotheoros*, i.e. three years after Huygens' death, and it is interesting to see how, in an attempt to lend Huygens' argument additional persuasive force, the translator develops the rhetoric of the more contained Latin original.

Huygens is unremitting in his endeavor to populate every single nook of the universe. Yet there is one major obstacle to his ambitious project: By the late seventeenth century, when he wrote his treatise, the quality of telescopes had been considerably improved. This led to the discovery of craters in the dark parts of the lunar surface (the so-called 'old spots' visible with the naked eye), which made it clear that these parts could not be covered by water as had previously been believed. In addition to the growing doubts about the existence of water on the Moon, it had come to be thought unlikely that the Moon had an atmosphere.[42] Without water and air, it was impossible to conceive of the Moon as being inhabited by forms of life even remotely similar to life on Earth. As a result, one of the principal arguments in support of the theory of the plurality of worlds was lost. Yet this was not all: An argument in favor of the possibility that lunar life could be radically different from the Earth would, first of all, have invalidated Huygens' argument for the basic uniformity of the different worlds in the cosmos. And since the Moon is the Earth's closest neighbor, it would also have invalidated the hypothesis formulated by both Fontenelle and

---

40 Shapiro, *Probability*, 239.

41 See Bono, "Role/Rule," 59-89. Bono argues that metaphorical aspects of language such as proliferation and dissemination of meaning are essential to understanding the dynamic of conceptual change in science because they ground complex scientific texts and discourses in other social, political, religious, or 'cultural' texts and discourses.

42 Huygens mentions both of these points in his treatise. Moreover, he objects to Kepler's fortress theory by arguing that the phenomena Kepler had interpreted as fortresses were far too big to have been built by lunar inhabitants. He thinks it possible that they were caused naturally. (*Cosmotheoros*, 131-32)

Huygens that the similarities between the Earth and other worlds were stronger the less distant these worlds were; it would have destroyed Fontenelle's 'universal vision'.

The translator's rhetorical device to bridge this dilemma works by referring to the planets and moons of our solar system in the metaphorical terms of a hierarchically structured society. Throughout Huygens' treatise, he refers to the satellites of Jupiter and Saturn as "attendants", "guards" and "gentlemen", and at first such references seem to be nothing but quaint modes of expression.[43] It is not until the second book, where the argument focuses more specifically on the different worlds in our solar system, that the purpose of these metaphorical expressions manifests itself in the elaboration of the metaphor: In comparing the various planets, the text refers to Jupiter and Saturn as the most "majestic" planets in our solar system; not only do they by far outsize all others, they also have several "attendants" that "wait upon them."[44] The argument then turns to the question of the existence of other inhabited worlds:

> And here one would think that when the Moon is so near us, and by the means of a Telescope may be so nicely and exactly observ'd; it should afford us matter for more probable Conjectures than any of the other remote Planets. But it is quite otherwise, and I can scarce find any thing to say of it, because *I have not a Planet of the same nature* before my eyes, as in all the primary ones I have. For they are *of the same kind* with our Earth; and seeing all the Actions, and every thing that is here, we may make a reasonable Conjecture at what we cannot see in those Worlds. But this we may venture to say, without fear, that *all the Attendants of Jupiter and Saturn are of the same nature with our Moon*, as going round them, and being carry'd with them round the Sun just as the Moon is with the Earth. Their Likeness reaches to other things too, as you'l see by and by. Therefore whatsoever we can with reason affirm or fancy of our Moon (and we may say a little of it) must be suppos'd with very little alteration to belong to *the Guards of Jupiter and Saturn*, as having *no reason to be at all inferior to that.*[45]

In order to salvage the notion of the habitableness of the planets, Huygens argues that the negative evidence in the case of the Moon is not representative for the real planets. He distinguishes between the Moon and the planets in terms of social class, arguing that planets and moons must be evaluated in accordance with their very different rank in the celestial hierarchy. If the evidence available makes it appear unlikely that the Moon is inhabited, this is likely to say something about the nature of those celestial bodies equal in rank, yet it should not be applied beyond.[46] The point the text leaves implied here is, of course, that there is no reason why the other planets, especially the ones which with their several moons are more 'majestic' than the Earth, should be at all inferior to it when it comes to inhabitants. By projecting the image of a hierarchically structured society onto the system of planets and moons, the translator plays

---

[43] *Ibid.*, 7, 12, and later, 112, 114, 128-29. The original text uses the more neutral metaphorical expression *comites* (i.e., companion) for the satellites. (*Oeuvres complètes*, vol. 21, 687 and 691)

[44] *Ibid.*, 112.

[45] *Ibid.* 128-29. Emphasis mine except in the names of the planets.

[46] It should also be noted that Huygens does not say anything yet about the negative quality of the evidence available, but, on the contrary, represents it as a valuable help for the characterization of the other moons in our solar system.

down the impact of the exclusively negative evidence available on the existence of other inhabited worlds: If 'guards' and 'attendants' do not show signs of life, this is not such a big deal – actually, it is entirely in keeping with their role: not to show character, but to serve their master. The natural order of the universe is here placed in correspondence with the social order of French society so as to imply that its hierarchic organization represents a universal standard.[47]

Once more, the use of metaphor here exceeds the purpose of mere illustration; by dictating structure, by furnishing causal connections, the metaphor gives narrative coherence to an argument about blank spaces. By clothing Huygens' argument in favor of the habitableness of other planets in a cultural discourse and by arguing in the terms of this discourse, the translator is, of course, above all trying to cover up argumentative straits with the help of a cheap rhetorical trick. However, to some extent, the metaphor of a social hierarchy also makes it possible to construct a system of the universe by offering a structure on which a hypothesis may be formed.

It is interesting to consider this metaphor of social hierarchies in the context of contemporary utopian literature, which typically showed a strong concern for issues of social hierarchy, though not as metaphor. Famous examples of utopian societies organized according to a strict hierarchic order are Tommaso Campanella's *La città del Sole*, Denis Veiras' *Histoire des Sévarambes* and the *Voyages et avantures de Jacques Massé* by Tyssot de Patot. While such accounts of utopian societies are often set in the Antipodes (e.g., Veiras and Patot), there are numerous utopian and distopian societies described in accounts of imaginary voyages to the Moon and other planets.[48] As Karl Guthke has observed, the existence of rational human beings on the Antipodes and the fervent debates about their existence before its confirmation were frequently used in support of the theory of the plurality of worlds inhabited by rational beings in the larger universe. The parallelism between the antipodal world and the putative worlds beyond the Earth was therefore well established.[49] Accordingly, Huygens' metaphor of social hierarchies buys into a discourse that was somehow in the air.[50]

---

47 Interesting in this context is also the sequel of Huygens' argument, in which he contemplates the idea of planting colonies on the moons: What is striking about his speculations is that they are made exclusively in connection with moons while no mention is made about the possibility of colonizing one of the planets. Here again, the distinction of social class comes into play: Territories whose owners are of socially equal (or higher) rank must be respected as rightfully owned while others are available on a first come first served basis. Pierre Borel makes a similar argument about social propriety and the way it corresponds to reason and the natural world when he rejects the claim that the innumerable heavenly bodies have been created for the Earth and its inhabitants alone: "car la raison naturelle nous dissuade assez de croire que les choses plus grandes servent aux petites, & que celles qui sont beaucoup plus nobles servent aux viles & de moindre consequence, n'est-il pas plus vray semblable que chaque globe fasse un monde, ou une terre particuliere". Cf. *Discours nouveau prouvant la pluralité des Mondes* (Geneva, 1657), 12.

See also Timothy Reiss' discussion of the "homology of the natural, social and rational world" and its link to the concept of poetic justice in seventeenth- and eighteenth-century debates about the ethical responsibility of poetry. (*Meaning*, 171)

48 See for example the more egalitarian society described by Francis Gentleman in *A Trip to the Moon*.

49 Cf. *Mythos, passim*.

50 There is a nice local example which shows that the metaphor of a social hierarchy representing a heliocentric cosmos in fact reverses a metaphorical representation that seems to have been very

At this point, I would like to return to the *Conversations*, where the political metaphor of neighboring states is projected beyond our solar system onto the many solar systems supposed to exist around the fixed stars. According to the Cartesian theory that Fontenelle adopts, the universe consists of innumerable vortices of different sizes, each with a sun at its center. The interaction between vortices is described in terms of neighboring states that, since each is constantly striving to expand, are "always in combat" (*se combattent toujours*) with each other. At the same time, they also cultivate their relations with their neighbors by sending them "emissaries" (*députés*) in the form of comets, which the narrator defines as planets that move close to the outer edges of their vortex and, completing their orbit, enter into the next vortex. Such 'emissaries' are "adorned with dazzling hair, or a venerable beard, or a majestic tail."[51] When they appear in our vortex, they take up their dwelling in its outermost region.

The Marquise translates this observation into another analogy with foreign politics: "'I understand,' she said. 'We don't allow them to enter the heart of our vortex with our planets; we receive them as the sultan of Turkey receives the ambassadors who are sent to him. He doesn't give them the honor of lodging in Constantinople, but only in a suburb of the city.'" To this the narrator adds: "'We also have this in common with the Ottomans ... that they receive ambassadors without sending any in return, and we don't send any of our planets to neighboring worlds.'"[52] The Marquise thinks we are proud not to do so; and when the narrator tells her that changing vortex is a highly unpleasant experience, she does not believe him, but thinks instead that the comets are going on a delightful and highly edifying journey, a kind of *grand tour* in which their inhabitants get to see the universe like tourists visiting other countries.

Fontenelle's playful use of political and social discourses in his metaphorical description of cosmic spectacles is particularly interesting for the way the levels to which the metaphors are applied keep changing and are mixed with each other: Accordingly, the same metaphors are projected onto the structure of our solar system and from there onto the potentially infinite arrangement of solar systems in the wider universe, only to return to their literal level in the reference to the inhabitants of another planet and their adventurous journey through space. As an effect of the application of the same metaphors to different levels and the reference to an object at one

popular at the time: At the *Schweizerisches Landesmuseum* in Zurich, there is a wooden chest dating from 1657 which represents the hierarchic order of the government of Zurich in terms of celestial constellations. The center-piece originally contained 160 discs representing the various governmental posts. The central disk, around which all others revolve, is assigned to the highest posts. Each disk is filled with the coat of arms of the people who held the respective posts in the period from 1490 right through 1798, the year in which the Swiss Confederacy was turned into the Helvetian Republic under Napoleon. Cf. Rudolf Schnyder, ed., *Zürcher Staatsaltertümer: Der Zürcher Staat im 17. Jahrhundert* (Bern: Paul Haupt, 1975), 14.

51 *Conversations*, 68. – "qui sont ornées, ou d'une chevelure éclatante, ou d'une barbe vénerable, ou d'une queuë majestueuse." (*Entretiens*, 145)

52 *Conversations*, 69. – "J'entends, dit-elle. Nous ne leur permettons pas d'entrer jusque dans le coeur de notre Tourbillon, et de se mêler avec nos Planetes, nous les recevons comme le Grand Seigneur reçoit les Ambassadeurs qu'on lui envoye. Il ne leur fait pas l'honneur de les loger à Constantinople, mais seulement dans un Fauxbourg de la Ville. Nous avons encore cela de commun avec les Ottomans, repris-je, qu'ils reçoivent des Ambassadeurs sans en renvoyer, et que nous ne renvoyons point de nos Planetes aux Mondes voisins." (*Entretiens*, 147)

time as a world and at another as a human individual, the metaphorical image starts to oscillate between its different meanings. It is in this oscillation that the metaphor carries its greatest potential for the imagination. To illustrate this effect in more detail and define its creative potential, I would like to turn to the analogy drawn between the concept of plural worlds in the universe and the human body as a system of innumerable smaller bodies. We will see that the metaphorical terms in which this analogy is presented directly integrate it into the representations of the larger cosmos as a social and political network, that the latter, in fact, owe their main significance to the metaphorical adaptation of this analogy.

There is no question that the introduction of the telescope to astronomy greatly helped to promote Copernican theory and the concept of plural worlds. Galileo's discoveries of the moons orbiting Jupiter, the phases of Venus, the earthlike nature of the Moon and a great number of stars never perceived before, not only offered empirical evidence for Copernicus' cosmology, but also invited speculations about the multiplication of this cosmological system in a much vaster, potentially even infinite universe and about the plurality of inhabited worlds in this universe. Yet while the evidence that the telescope supplied may have increased the likelihood of such speculations, it was in no way sufficient to carry them beyond the merely speculative. At the same time, new worlds were being discovered much closer to home: The microscope, invented shortly after the telescope, revealed to the human eye new forms of life infinitesimally smaller than any that had been known so far. Both of these instruments seemed to imply that yet smaller and remoter worlds would reveal themselves endlessly as the techniques of lens-grinding continued to be further refined.

It is now interesting to see how the discoveries made with the microscope are exploited to the advantage of the theory of plural worlds in the cosmos by being presented in metaphorical terms. After expressing his disbelief regarding the Earth's uniqueness as an inhabited world, the narrator of the *Conversations* explains to the Marquise that, of the creatures inhabiting the Earth, those visible to the naked eye are by no means all that exist: "We see from the elephant down to the mite; there our sight ends. But beyond the mite an infinite multitude of animals begins for which the mite is an elephant, and which can't be perceived with ordinary eyesight".[53] What is striking about this observation is that the narrator refers to both the visible and the invisible world in the same terms. The order manifest from the largest (elephant) to the smallest of visible creatures (mite) is simply repeated in the order among the invisible creatures so as to suggest that the latter order is a mere replica of the former. Accordingly, the system of the world, as human beings experience it, is here projected in the opposite direction.

Cyrano's *Lune*, which Fontenelle was obviously familiar with, contains an elaborate account of the microscopic world that works according to the same principle, with the difference that, instead of Fontenelle's paratactic structure, the smaller system is here contained in the larger one: The microscopic world is contained in the human body, thus turning this body into a world. Moreover, Cyrano, like Fontenelle, elegantly com-

[53] *Conversations*, 44-45. – "Nous voyons depuis l'Elephant, jusqu'au Ciron, là finit notre vûë; mais au Ciron commence une multitude infinie d'Animaux, dont il est l'Elephant, et que nos yeux ne sçauroient appercevoir sans secours." (*Entretiens*, 92-93)

bines the microcosmic and the cosmic perspective: The description of the world in miniature forms part of a conversation between Dyrcona and a philosopher of the Moon in which the philosopher tells Dyrcona about his theory that the universe consists of innumerable worlds of different scales locked into each other. This is how he describes the world within Dyrcona's body:

> For do you find it hard to believe that a louse should take your body for a world, or that, when one of them travels from one of your ears to the other, his friends should say that he has voyaged to the ends of the earth, or that he has journeyed from pole to pole? Why, doubtless this tiny people take your hair for the forests of their country, your pores full of sweat for springs, your pimples for lakes and ponds, your abscesses for seas, your streaming nose for a flood; and when you comb your hair backwards and forwards they think this is the ebb and flow of the ocean tides.
>
> Does not the itch prove my point? The mite which produces it is surely none other than one of these little animals, which has broken away from civil society and set itself up as a tyrant in its own country. If you ask me how such creatures come to be larger than the rest of their imperceptible fellows, I will ask you why elephants are bigger than us and Irishmen bigger than Spaniards. As for your blister and your scab, whose origins are unknown to you, they must either result from the rotting carcasses of enemies slaughtered by these little giants, or else a plague (caused by the lack of foodstuffs upon which the rebels have gorged themselves) has left behind heaps of corpses, or else the tyrant has driven away all his neighbours, whose bodies stopped up the pores in our own, thus making a passage for the phlegm, which then escapes from the bloodstream and becomes corrupted.[54]

> Car dites-moy, je vous prie, est-il mal aisé à croire qu'un pou prenne vostre corps pour un Monde, et que quand quelqu'un d'eux a voyagé depuis l'une de vos oreilles jusques à l'autre, ses compagnons disent de luy qu'il a voyagé aux deux bouts du Monde ou qu'il a couru de l'un à l'autre Pôle? Oüy, sans doubte, ce petit peuple prend vostre poil pour les forests de son païs, les pores pleins de pituitte pour des fontaines, les bubes et les cirons pour des lacs et des estangs, les apostumes pour des mers, les fluxions pour des déluges; et quand vous vous peignés en devant et en arrière, ilz prennent cette agitation pour le flux et le reflux de l'Océan. La démangeaison ne prouve-t-elle pas mon dire? Ce ciron qui la produit, est-ce autre chose qu'un de ces petits animaux qui s'est dépris de la société civille pour s'establir tiran de son païs? Si vous me demandés d'où vient qu'ilz sont plus grands que ces autres petits imperceptibles, je vous demande pourquoy les éléphans sont plus grands que nous, et les Hibernois que les Espagnols? Quant à cette ampoulle et cette crouste dont vous ignorés la cause, il fault qu'elles arrivent, ou par la corruption des charognes de leurs ennemis que ces petits Géans ont massacrés, ou que la peste produite par la nécessité des alimens dont les séditieux se sont gorgez ayt laissé pourrir parmy la campagne des monceaux de cadavres, ou que ce tiran, après avoir tout autour de soy chassé ses compagnons qui de leurs corps bouchoient les pores du nostre, ayt donné passage à la pituitte, laquelle, estant extravasée hors la sphère de la circulation de nostre sang, s'est corrompüe.[55]

---

54 *Other Worlds*, 75-76.
55 *Lune*, 71-72.

In the *Iter lunare*, David Russen's commented summary of Cyrano's lunar voyage, this passage is offered with some interesting elaborations: Russen refers to the "Elephantine Bulk of the Skipping Flea, in shape like an Elephant and of a Bulk as disproportionable to those little Animals as the Elephant is to other Beasts" and compares the louse to our rhinoceros. Moreover, in an ingenious fusion between the microscopic and the cosmic perspective, he compares the impression the movements the human body create on its tiny inhabitants to the scene the revolutions of celestial bodies offer to human beings.[56]

What is remarkable about this little exposition is that the microscopic world is described as the exact replica of the world of human beings. The human body becomes a world in itself, landscaped with forests, springs and ponds, in no way distinguishable from how the Earth appears to a human being. It is populated by creatures of many different kinds and sizes, but basically corresponding to the variety of species found in the larger world. Rather than the estrangement of the familiar that looking through a microscope typically entails, Cyrano emphasizes the essential sameness of this miniature world once the perspective is naturalized. Not only are pimples, blisters and scabs turned into pleasant landscapes, even the life of microbes is humanized to the last detail: We find friendship, revolution, war, societies ravaged by plagues and, in David Russen's little elaboration of the scene, even astronomy.[57]

With the doubling of worlds and the change of perspective it presupposes, the distinction between the literal and the metaphorical starts to blur, and it is no longer clear which of the two worlds serves as the tenor and which as the vehicle of the metaphor. Yet this is only the beginning of the confusion: Both Fontenelle and Cyrano set their descriptions of the microscopic world as miniature replicas of the world of human beings in the immediate context of the theory of the plurality of worlds in the larger cosmos.[58] Fontenelle's metaphorical representations of the cosmos, or parts of it, as an enlarged version of the social and political structures in which human beings organize their relationships are therefore only a logical continuation of the pattern

---

56 *Iter lunare: or a Voyage to the Moon* (London: 1707), 132-33.

57 There is an analogous passage in Sorel's *Histoire comique de Francion* that is worth quoting since it seems to have served Cyrano as a source: "Sachez que si le monde nous semble grand, notre corps ne le semble pas moins à un pou ou à un ciron. Il y trouve ses régions et ses cités. Or il n'y a si petit corps qui ne puisse être divisé en des parties innombrables; tellement qu'il ne peut faire que dedans ou dessus un ciron, il y ait encore d'autres animaux plus petits qui vivent là comme dans un bien spacieux monde". (Quoted by Calame in *Entretiens*, 93, note 1)

58 In an earlier conversation between the Viceroy of Canada and Dyrcona, the latter combines the microcosmic and the cosmic dimension even more emphatically. He also uses the political metaphor of the solar system as a monarchy: "Firstly, it is common sense to believe that the sun has its place at the centre of the universe, since all the bodies which exist in nature have need of this fundamental source of heat. It dwells in the heart of the kingdom so that it may swiftly satisfy the needs of each region. The first cause of all life is situated at the very centre of all bodies in order to function with equity and with greater ease. Wise nature has located the genital organs in man after the same fashion, pips at the centre of apples and stones in the middle of their fruit. The onion likewise protects, under the shield of the hundred skins which surround it, the precious seed whence ten million others must draw their essence. *For the apple is a little universe in itself, wherein the pip, warmer than the other parts, is a sun giving off the heat which preserves the rest of its sphere, and the onion seed, according to this theory, is the little sun of its own little cosmos, which warms and nourishes the vegetative salts of this little body.*" (*Lune*, 8, emphasis mine)

established between the human and the microscopic world. This pattern also entails a change of perspective in the opposite direction, which turns human beings into mites, so that their lakes metamorphose into some giant creature's pimples and scabs and all that is left of a pleasant grove is a sparse cluster of corporeal hair. Though his terms are not as blatant, this is basically what we find when Fontenelle likens the Earth with its atmosphere to a silkworm's cocoon,

> made of very compact silk, but ... covered by a very light and soft down. In the same way the Earth, which is solid enough, is covered to a height of twenty leagues more or less by a kind of down, which is the air, and the whole cocoon turns at the same time. Beyond the air is the celestial matter, incomparably more pure, more subtle, and more agitated than it is.[59]

> [Les coques de ver à soie] sont d'une soye fort serrée, mais elles sont couvertes d'un certain duvet fort leger et fort lâche. C'est ainsi que la Terre, qui est assés solide, est couverte depuis sa surface jusqu'à une certaine hauteur, d'une espece de duvet, qui est l'Air, et toute la Coque de Ver à Soye tourne en même temps. Au-delà de l'Air est la matiere céleste, incomparablement plus pure, plus subtile, et même plus agitée qu'il n'est.[60]

Upon the Marquise's remonstrance that he is unduly trivializing the great deeds performed on Earth, Fontenelle playfully continues

> Yes, ... and all the while Nature, who takes no notice of these separate little stirrings, carries us all together in a general movement, and plays with this little ball.[61]

> Oüi, ... et pendant ce tems-là la Nature qui n'entre point en connoissance de tous ces petits mouvemens particuliers, nous emporte tous ensemble d'un mouvement general, et se joüe de la petite boule.[62]

– thus turning the cocoon into Dame Nature's plaything while the ball with its twofold rotation serves him as a perfect illustration for the Earth's double motion round its axis and round the Sun.

By representing his system of the cosmos as a continuum of smaller worlds locked into greater ones like Russian dolls, Cyrano adds stringency to the pattern: If, within the narrow confines of the human body, the microscope reveals an entire world, what better image could there be for the concept of a cosmic plurality of worlds than the image of a human society, itself structured into individual political bodies – i.e., the image of a boundlessly expandable Leviathan that in each of his forms represents both a natural and a political body? In Cyrano's system of the cosmos, each part of creation is simultaneously part and whole, depending on the perspective taken. The system is the product of a multiple as well as reversible synecdoche where the part not only implies the whole by virtue of being a part of it and therefore logically connected with

---

59 *Conversations*, 21-22.

60 *Entretiens*, 42. Addition mine.

61 *Conversations*, 22.

62 *Entretiens*, 43.

it, but where the part is a smaller replica of the whole by being represented in its image. As a result, even though the synecdochic structure contains several levels, any part can be used to represent any whole and vice versa – the representation works in both directions, as *pars pro toto* and *totum pro parte.*[63]

The pattern that Cyrano and Fontenelle apply to the cosmos derives from the traditional concept of microcosmic and macrocosmic relations; it holds that the universe has an architectonic structure exhibiting patterns of correspondence among its various levels and individual parts. However, the finite structure of relations of the traditional concept, for which man is the ultimate referent, is here broken up and replaced by infinite relativity. This leads us back to the metaphoric: Besides serving as vehicle for the metaphorical representation of both the microcosmic as well as the (macro)cosmic world, the world of human beings also constitutes the mediating link between the two, on which the structural pattern of the cosmos depends. Yet because the metaphor works in two directions – the human body at one time serving as metaphor for the world as it is experienced from the microbe's perspective, at another for the microbe itself in a much larger world – the unidirectionality of its meaning is disturbed, and it is no longer clear whether the world of human beings is to be taken as the tenor or as the vehicle of the metaphorical relation: The metaphor is unhinged and assumes a chiasmic structure.[64]

This ambiguity of perspectives offers the greatest potential of this metaphor: When the world familiar to human beings is metaphorically projected onto the larger and smaller cosmos to form a pattern of potentially infinite repetition, this obviously serves to make the universe picturable in both its unfamiliar and unknown dimensions. It moreover offers a recipe for an infinite universe by conceiving of it in parts and adding on to them.[65] At the same time, the polyvalence of the metaphor and the changes of perspective it entails deprive the familiar world of its status as stable point of reference and thus question the legitimacy of its pictorial representation. The instability of the image in fact reminds us that the image is metaphorical. Further, the continual shift of perspectives, by which the familiar world switches from being the metaphorical image to being the object metaphorically illustrated, to some extent rids the representation of its images, leaving behind only the system of relations that underlies it, which remains

---

[63] For a discussion of the importance of synecdoche in Copernican rhetoric, see Hallyn, *Poetic Structure*, 105-25. Hallyn argues that by claiming that there are multiple gravitational centers in the universe, Copernicus attributes to the parts what hitherto only applied to the whole universe. His physical theory is thus the result of a synecdoche.

[64] This chiasmic structure is well expressed in the Fontenelle's reference to the world seen through the microscope: "We see from the *elephant* down to the *mite*; there our sight ends. But beyond the *mite* an infinite multitude of animals begins for which the mite is an *elephant*, and which can't be perceived with ordinary eyesight". (*Conversations*, 44-45, emphasis mine)

[65] Compare Dyrcona's advice to the Viceroy, who tells the traveller that he cannot conceive the infinite: "'Why! tell me,' I countered, 'do you understand the nothingness beyond the end of space? Not at all! For when you think about this nothingness, at the very least you imagine it like wind, or like air – and that is something. But even if you do not understand infinity as a whole you may at least conceive of it in parts, since it is not difficult to picture, beyond the earth, air, and fire which we can see, more air and more earth. And infinity is really nothing but a limitless texture of all that.'" (*Lune*, 11)

the same.[66] In this way, the ambiguity has a liberating function, for it permits the imagination to overcome its natural limitations. This liberation can never last, however. It is only for the brief moment in which the stability of the point of reference is shaken, the moment when meanings are oscillating, that our imagination is freed from the inhibiting specificity of images and achieves its maximal creative power.

It is useful to refer here briefly to Paul Ricoeur's discussion of the central role of imagination in the reception of metaphor. Ricoeur emphasizes the creative moment in metaphor by arguing that metaphor cannot be identified with the clash it produces on the literal level, but that it is rather the resolution of this clash. He then argues that this resolution never merges into identity. Instead, "metaphor is that place in discourse where ... the identity and the difference do not melt together but confront each other."[67] "The insight into likeness is the perception of the conflict between the previous incompatibility and the new compatibility. 'Remoteness' is preserved within 'proximity'."[68] This tension is where figurative (or, as I have been calling it, virtual) meaning emerges. Metaphor remains productive rather than merely reproductive as long as it maintains this tension, that is to say, as long as it is perceived as metaphoric.

If the play with perspectives in Cyrano's system of the cosmos offers a good example of a moment when the imagination transcends the scope of its images and approaches the status of an abstract structure for thought, the narrative flow of the *Conversations* as a whole, with its constantly shifting perspectives and changing propositions, marks a similar attempt to free the imagination from the inhibiting specificity of images. The theme of mobility is of central importance in the *Conversations.* "The narrator's own position is mobile: His constant shifts of argumentation are paralleled by his emphasis on relations and reversibilities and by the literal multiplication of viewpoints through the notion of the interplanetary voyage" in the conversation of the fourth evening.[69]

In his nightly conversations with the Marquise, the narrator repeatedly conjures up conceivable forms of life on the Moon and the planets only to reject the possibility of their existence again by questioning a basic premise of the chain of hypotheses; in this way, he exposes his theories as a mere *jeu d'esprit.* The best example of this is at the beginning of the third conversation, where, to the Marquise's great disappointment, the narrator suddenly dismisses all his earlier theories regarding forms of lunar life because, to all appearances, the Moon has neither water nor an atmosphere. To satisfy the Marquise's wishes, he then proposes other means of sustenance that would allow the existence of life. At the same time, he reminds her that she "should never give more than half [her] mind to beliefs of this sort, and keep the other half free so that the

---

66 James Bono offers a fascinating discussion of Mersenne's idea of how the grammar of any language (the abstract concept of grammar) can serve man as a blueprint for the understanding of nature. Language according to Mersenne is wholly arbitrary. Yet while language cannot give man insight into the essence of nature, its very structure "provides us with a tableau of the world in its ever-changing configurations of relationships." The "Art of Grammar" "captures something real about nature" by revealing it as a network of things. (*Word of God*, 262-63)

67 *The Rule of Metaphor* (Toronto: University of Toronto Press, 1977), 190 and 199.

68 "The Metaphorical Process as Cognition, Imagination, and Feeling," in Mark Johnson, ed., *Philosophical Perspectives on Metaphor* (Minneapolis: University of Minnesota Press), 234.

69 Howells, "Principle of Mobility," 129.

contrary can be admitted if it's necessary."[70] As Howells aptly observes, "[t]he multiplication of models is itself heuristic. It illustrates the notion of plural possibilities. If each is far less than rigorous so much the better. The purpose is to suggest new ways of thinking. If they contradict each other, so much the better. The purpose is to refuse closure."[71] In contrast to the typically linear movement of scientific argumentation, the *Conversations* have an *ironic* structure: The hypotheses established are not confirmed or dismissed in a progressive order, but they keep being questioned through other hypotheses, thus lending the theories proposed an ironic circularity.

The mobility of argumentation in the *Conversations* seems to suggest that the real creative freedom lies in the continual act of making and unmaking worlds, at the point where one hypothesis is dismissed and exchanged for a new one, thus opening – or rather hinting at – a *potential* space for creation between the lines of any given narrative. The only chance for the imagination to escape the eternal cycle of reproduced images lies in the unimaginable: in ridding imagination of its visual quality, of the images that compose it. What needs to be traced is the *substans* (in the Aristotelian sense) beyond the image, the blue-print or skeleton on which the visual properties of the image rest, the order of its composition. To unfold the mere potential of creation, we need to discard the recycled image. While this is actually impracticable, it is most nearly achieved in the oscillation between perspectives and the turbulent rise and fall of hypotheses. Rather than in the creation of positive images, the imagination is most creative in the freedom to combine, take apart, restructure and recombine. Accordingly, the narrator once observes to the Marquise, who is perplexed by the lighthearted way in which he speculates on conceivable forms of life in the cosmos: "This 'Why not?' has a power which allows it to populate everything."[72]

---

[70] *Conversations*, 37. Adaptation mine. – "il faut ne donner que la moitié de son esprit aux choses de cette espece que l'on croit, et en reserver une autre moitié libre, où le contraire puisse être admis, s'il est besoin." (*Entretiens*, 76-77)

[71] "Principle of Mobility," 133. Adaptation mine.

[72] *Conversations*, 44. – "ce *pourquoi non?* a une vertue qui peuplera tout." (*Entretiens*, 91)

6

Representing the Unimaginability of the Imaginable: Italo Calvino's *Castello dei destini incrociati* and *Le città invisibili*

In the preceding chapters we have been analyzing texts where the unimaginable and hypothetical world beyond the limits of knowledge and sensual experience was made imaginable through metaphor. By representing it in metaphorical images, the unimaginable was made conceivable while, at the same time, leaving room for the unprecedented. So far, what I am referring to here as the 'unimaginable' was conditioned by the dependence of the imagination on sensory experience and the resulting impossibility of imagining worlds fundamentally different from the terrestrial world, i.e., the impossibility of imagining genuine alternatives rather than mere permutations of common experiences. Unimaginability was, as it were, a metaphysical problem and, therefore, a problem of the collective rather than the individual mind. In the last chapter, moreover, we have seen how metaphors came to serve as frameworks of relations by which a cosmological structure could be made conceivable even in its potentially infinite extension. The concept of infinity, which is both beyond imagination and beyond representation, was thus expressed with a simple metaphorical image by changing the levels of its application and letting it oscillate between them.

Although in this final chapter we are going to leave such cosmic exercises behind us to turn to more topical ones, the issues we are going to discuss in connection with the imagination will in many ways be similar to those of the earlier chapters. Accordingly, we will still be focusing on what is beyond imagination. Yet the notion of unimaginability will now be conditioned not by the limitations of the collective human imagination, but by the boundless multiplicity of collective experience and collective imaginability – i.e., the combined experiences of every individual consciousness (past, present and future) as well as the potential imaginative creations based on these diverse experiences. The limitation of the individual imagination to a realm of subjective experience, combined with the potential multiplicity of what is imaginable based on this experience, conveys a sense of the boundless diversity of the subjective world experience of different individuals in the past, the present and the future as well as the infinite multiplicity of what is, has been and will be imaginable based on all these diverse experiences. Moreover, it conveys a sense of the incompatibility of different imagining minds among each other. The collective experience of the world and what is imaginable with this experience thus inevitably remains unimaginable to the individual.

In addition to the problem of the impossibility for the individual of imagining collective world experience as well as what is, has been and will be collectively imaginable based on this experience, there is the even greater problem of the *representation* of experience and the imaginable (both individual and collective) – not just its representation in the mind, but its literary representation and narrative communicability. In "Visibility," one of his *Six Memos for the Next Millennium*, Italo Calvino refers to the artist's imagination as a "a repertory of what is potential, what is hypothetical", "a world of potentialities that no work will succeed in realizing."[1] Later, he describes each individual being as "a combinatoria of experiences, information, books we have read, things imagined[.] Each life is an encyclopedia, a library, an inventory of objects, a series of styles, and everything can be constantly shuffled and reordered in every way conceivable."[2] Yet if each individual constitutes its own encyclopedia of experiences, exchanges and relations, this encyclopedia is also connected to innumerable different encyclopedias opening themselves up around each person and thing. Clearly, the idea of total representation becomes more and more hopeless.

In "Multiplicity," the last essay of the *Memos*, Calvino dedicates himself to the idea of encyclopedic writing. Observing that several writers of the twentieth century have been obsessed with the task of representing the world as an infinitely complex tangle of relations, he claims that this idea is characteristic of the literature in this century, particularly the novel. In fact, he offers a definition of the contemporary novel "as an encyclopedia, as a method of knowledge, and above all as a network of connections between the events, the people, and the things of the world." Accordingly, the contemporary novel is here defined as a framework of relations that enables us to conceive the world as an order. Alongside writers like James Joyce, Marcel Proust and Robert Musil, Calvino focuses especially on Carlo Emilio Gadda and his lifelong endeavor "to represent the world as a knot, a tangled skein of yarn; to represent it without in the least diminishing the inextricable complexity."

> In these brief pieces, as in each episode in one of Gadda's novels, the least thing is seen as the center of a network of relationships that the writer cannot restrain himself from following, multiplying the details so that his descriptions and digressions become infinite. Whatever the starting point, the matter in hand spreads out and out, encompassing ever vaster horizons, and *if it were permitted* to go on further and further in every direction, it would end by embracing the entire universe.[3]

The conditional is crucial for, of course, the literary text has to come to an end and can never 'embrace the whole universe'. The same paradox is implied when Calvino argues that "[w]hat tends to emerge from the great novels of the twentieth century is the idea of an *open* encyclopedia, an adjective that certainly contradicts the noun *encyclopedia*, which etymologically implies an attempt to exhaust knowledge of the world

---

1 Cf. *Six Memos for the Next Millennium*. The Charles Eliot Norton Lectures 1985-86 (Cambridge, MA: Harvard University Press, 1988), 91 and 97. All quotations from this text refer to this edition. Henceforth cited as *Memos*. Calvino wrote these lectures in English. On account of his unexpected death, only five of the six memos originally planned were completed.

2 *Ibid.*, 124. Adaptation mine.

3 *Ibid.*, 105-7. Emphasis mine.

by enclosing it in a circle."[4] The idea of openness also contradicts the form of the literary work itself which, no matter to what dimensions it is extended, must finally come to an end. This is why, with the exception of shorter texts that revolve around a closely defined topic, all of Gadda's novels could be said to have remained unfinished.[5]

Finally, the book that tries to be an open encyclopedia of the world, no matter how exhaustive its pose, is of course always limited to the closed encyclopedia of knowledge, experience and imagination that is its author. For Bouvard and Pécuchet, the two protagonists of Flaubert's eponymous novel (according to Calvino, "the most encyclopedic novel ever written"), Flaubert is ironically transformed into a universal encyclopedia. In order to feed the adventures of his two naive, but voracious readers chapter for chapter, he had plunged himself into an ocean of books and drowned: the book remained unfinished.[6] How then can any book include what it might have said, but does not say? How can it break out of the solipsistic world of its author and become an 'open encyclopedia'? How can it represent the world of collective experience unavailable to its author? And how can it represent the world beyond any personal experience, give voice to the brute material facticity of things, what Jean-Paul Sartre has called being-in-itself (*être en soi*)?[7] On the last page of his *Memos*, Calvino expresses the following desire:

> Think what it would be to have a work conceived from outside the *self*, a work that would let us escape the limited perspective of the individual ego, not only to enter into selves like our own but to give speech to that which has no language, to the bird perching on the edge of the gutter, to the tree in spring and the tree in fall, to stone, to cement, to plastic ...[8]

In what follows, I will show how in his own literary works Calvino tried to find techniques that would allow him to do justice to the infinite universe of potentiality that either lies beyond the scope of actual representation in his book or simply beyond the scope of his imagination and that of any individual. The books I have chosen for this purpose are *Il castello dei destini incrociati* (*The Castle of Crossed Destinies*) and *Le città invisibili* (*Invisible Cities*), two books one might read as parables on literary representation and the imagination: By contrasting the representable and the represented with all that is lacking, they draw attention to the insurmountable conflict between linguistic representation and sensible experience and between the sensible experience and imagination of the authorial consciousness and a world of experience, imagination and material reality beyond it.

---

4 *Ibid.*, 116. Adaptation mine.

5 One such shorter text where, according to Calvino, Gadda's "encyclopedism" leads to a finished composition is a recipe for a "risotto alla milanese". (*Ibid.*, 106)

6 *Ibid.*, 114-15.

7 For Sartre's ontological distinction between being-in-itself (*en soi*) and being-for-itself (*pour soi*), see *L'être et le néant; essai d'ontologie phénoménologique* (Paris: Gallimard, 1957). While being-in-itself designates the brute material facticity of things, being-for-itself refers to the power of consciousness to negate the *en soi* and thereby achieve a distance or freedom from it – that is, an existence *pour soi*. But this consciousness is only free because it is a 'nothingness', a negation of the external world of the *en soi*.

8 *Memos*, 124.

My choice of Calvino is also to some extent an apology for his work, as Calvino has been admired, but also strongly criticized for his crystalline artifacts. Critics have censured his work for being too aesthetically clean and self-contained to relate to the real world; they have reproached Calvino for not taking into account the contingency, the chaotic disorder and the ugliness that rule this world. Most recently, Carla Benedetti has launched a harsh polemic against Calvino's work and its almost instant canonization by literary critics, setting it in contrast with the work of Pier Paolo Pasolini and the refusal customary among critics to accept especially Pasolini's later writings as 'genuine literature'.[9] Her objections to Calvino's work that concern us here are that his novels are limited to playing games within a purely literary world, which as a strictly autoreferential system shuts itself off from the world outside. In her view, Calvino's stylized literary productions reflect a form of escapism in so far as they turn away from the real world with its anxieties and its painful alterity;[10] they altogether lack involvement with the world and make a necrophiliac use of language by reducing it to ironic quotation.[11]

Benedetti argues that even in his last works, *Palomar* and the *Memos*, where Calvino changed direction and turned to the physical world beyond language, writing continues to be a game played within the closed boundaries of its own autoreferential system. While in his earlier writings Calvino played with narrative structures, he is now engaged in "playing the natural scientist who observes the physical world in order to describe it with as much exactness as possible. It is a game ... which, within its rules, contemplates a world to describe."[12] Benedetti criticizes Calvino's emphasis on exactitude, one of the six values advocated in the *Memos*, arguing that his idea of literature, as much as it may appear to appeal to a universal value, is really based on a drastic restriction: It excludes possible values in favor of a single one. It reduces literature to a constative speech act rather than making it performative. Benedetti therefore challenges what she polemically claims to be Calvino's "ideology of the constative" with the question: "Why should literature force itself to describe the world with precision rather than to change it, to open new perspectives, to make visible what is invisible, and other things of the kind?"[13] From her confrontation between the two writers, it is clear that all the shortcomings she identifies in Calvino are compensated by Pasolini.

I have summarized Benedetti's argument at some length here because I want to show that Calvino does not fit quite so neatly into the compartment she has devised for him. Indeed, the reason he does not is what makes his work interesting in our context. We will see that a central concern of the texts we are going to discuss is not so much the autoreferentiality of literature as the autoreferentiality of the mind, its solipsism. In *Invisible Cities*, in particular, there is a strong concern with the relation between mental concepts and the reality they are taken to refer to, their incompatibility with this reality; a concern with the subjectivity of mental concepts and the incompatibility between different mental worlds.

9 *Pasolini contro Calvino: Per una letteratura impura* (Torino: Bollati Boringhieri, 1998).

10 *Ibid.*, 58.

11 *Ibid.*, 53.

12 *Ibid.*, 125. Translation mine.

13 *Ibid.*, 127. Translation mine.

At the same time, it is important to emphasize that what Benedetti calls Calvino's idea of literature is not limited to the notion of 'exactness'. Calvino is just as much concerned with the notion of 'multiplicity', the theme of another chapter in the *Memos*, which Benedetti completely disregards. Accordingly, in both the *Castle of Crossed Destinies* and the *Invisible Cities*, Calvino is relentless in his attempt to create a text that breaks out of its own limits and out of the solipsistic perspective of an individual mind (his own, that of the fictitious characters in the narrative and that of the reader), so as to do justice to the alterity of existence and of thought and, in this way, make room for improvement and change. For change begins with the constant re-evaluation of the world we live in, of our position in this world and of our conception of it, as well as with accepting both the legitimacy and the possibility of alternative conceptions of this world. At the very end of *Invisible Cities*, Marco Polo explains to Kublai Khan:

> The inferno of the living is not something that will be; if there is one, it is what is already here, the inferno where we live every day, that we form by being together. There are two ways to escape suffering it. The first is easy for many: accept the inferno and become such a part of it that you can no longer see it. The second is risky and demands constant vigilance and apprehension: seek and learn to recognize who and what, in the midst of the inferno, are not inferno, then make them endure, give them space.[14]

> – L'inferno dei viventi non è qualcosa che sarà; se ce n'è uno, è quello che è già qui, l'inferno che abitiamo tutti i giorni, che formiamo stando insieme. Due modi ci sono per non soffrirne. Il primo riesce facile a molti: accettare l'inferno e diventarne parte fino al punto di non vederlo più. Il secondo è rischioso ed esige attenzione e apprendimento continui: cercare e saper riconoscere chi e cosa, in mezzo all'inferno, non è inferno, e farlo durare, e dargli spazio.[15]

Calvino is not just interested in the arbitrary link between words and things; he is just as much interested in the arbitrariness of conceptual frameworks and in their multiplicity. In the texts we are now turning to, this multiplicity of perspectives and the existence beyond them manifests itself at the point where representation fails, where the ostensibly fragmentary nature of what is represented points to an inexhaustible potential of possibilities. It manifests itself through absences and in the potential they circumscribe. Calvino's texts are not closed; they refuse closure. In my discussion of Calvino's two most crystalline works, I hope to show that while these works are obviously artificial, they are not self-contained; instead, they are meant to refract the world in the infinite potential they manifestly leave implied. Moreover, Calvino's crystals also include cracks and blind spots which show that their perfection is only illusory.

The hidden potential in Calvino's texts manifests itself through the following techniques: abstraction and its contraposition to concrete detail; repeated attestations that an adequate description is impossible as well as guidelines about what such a descrip-

---

14 *Invisible Cities*, trans. William Weaver (New York: Harcourt Brace Jovanovich, 1972), 165. All quotations are taken from this edition, henceforth cited as *Cities*.

15 *Le città invisibili* (Torino: Einaudi, 1972), 170. Italics I.C. All quotations in Italian are taken from this edition, henceforth cited as *Città*.

tion would have to include; contradiction; ambiguity; changes of signification through changing contexts; the switching of roles between narrator and listener/observer/reader to emphasize that the narrative offered is just one of an infinite number of possible ones that remain implied; and, finally, the open, fragmentary character of the text, which, rather than just denoting failure, marks the existence of something beyond what is represented in the text.

I will begin with the *Castle of Crossed Destinies* since it is the simpler text of the two. Let me briefly recall the particular narrative situation presented in this book: Eleven characters meet in a castle in the midst of a dense and mysterious forest. After the terrors experienced on their journey across the forest, they feel a great urge to relate their adventures to the others, but they soon realize that the terror has left them bereft of the faculty of speech. To overcome the general aphasia, they use a deck of tarot cards as a tool for communication. One character after the other tells his or her story by laying cards in a double row, each row attached to that of a previous 'narrator'. Though all the 'narrators' have the same number of cards at their disposal, the first narrator obviously enjoys the greatest liberty to choose his cards and determine the order in which he wants to lay them. In contrast, the others are forced to adopt cards already used in the stories of other 'narrators'. This limitation gradually increases as the network of tarot cards grows denser. At some point, all the cards have been used up and have come to form a crystalline structure. Indeed, if it were not for the oblong shape of the cards, they would form a square approaching the form of a ring, i.e., an en*cyclo*pedia. The remaining characters have to tell their stories by using the cards that constitute a previous narration in inverted order. While one character lays his or her cards, the others are intent on observing the miens and gestures that accompany the narration so as to guess the story. The last of the characters to arrive at the castle presents his interpretation of each narration to the reader.

In considering this crossword of images, I would like to recall Calvino's definition of the contemporary novel as a 'network of connections between the events, the people, and the things of the world': as it happens, the network of connections is just what is missing here. All there is are the images on the cards, which function as symbolic and therefore ambiguous representations of the events, persons and things of the world that need to be linked together in a narrative. And while such a narrative is offered, the role of the narrator is split to the effect that the narrative the text presents is just one of an unlimited number of possible ones: The characters laying cards, who act as narrators on the first level, can only determine the order in which they want to arrange their cards (and this also only partly, as some of them have to reuse the cards used by earlier 'narrators'); the interpretation of the cards and the action that connects them must be furnished by the onlookers. They may be influenced by the miens and gestures with which the card-layers accompany their ritual and with the 'style' of each individual card-layer, but these, too, are subject to interpretation.

The narrator-interpreter who presents his version of the various narratives in the book is therefore constantly torn between different potential meanings he imagines to recognize in the images on the cards; he has to decide which interpretation makes most sense in the context. Part of the uncertainty as to what the image on a tarot card could have been intended to represent lies in whether its image should be read literally or figuratively. The following passage offers a good illustration of the difficulties the narrator of the book faces:

> The interpretation of this passage in the tale [the tarot called *The Sun*] was not easy. It could simply mean "it was a fine sunny day," and in this case our narrator was wasting his cards telling us inessential details. But perhaps the allegorical significance of the picture was less to be dwelt on than its literal meaning: a half-naked child had been seen running in the neighborhood of the castle where the wedding was being celebrated, and the groom had deserted the banquet to pursue that urchin. ...
>
> We hoped these questions would be answered by the following card, and when we saw it was *Justice* we were convinced that in this Arcanum – which showed not only a woman with sword and scales, as in ordinary tarot packs, but also, in the background (or, according to how you looked at it, in a lunette over the main figure) a mounted warrior (or an Amazon?), in armor, moving to the attack – there was contained one of the chapters richest in adventures of our tale. We could only venture some guesses.[16]

> L'interpretazione di questo passo del racconto [il tarocco chiamato *Il Sole*] non era facile: poteva voler dire semplicemente 'era una bella giornata di sole' e in questo caso il nostro narratore sprecava le sue carte per riferirci particolari inessenziali. Forse più che sul significato allegorico della figura conveniva soffermarsi su quello letterale: un bambino seminudo era stato visto correre nelle vicinanze del castello etc. ... Questi interrogativi speravamo ci fossero chiariti dalla carta seguente, e quando vedemmo che essa era *La Giustizia* ci convincemmo che in quest'arcano – il quale non mostrava soltanto, come nei comuni mazzi di tarocchi, una donna con la spada e la bilancia, ma anche, sullo sfondo (oppure, a seconda di come si guardava, su di una lunetta sovrastante la figura principale) un guerriero a cavallo (o un'amazzone?) in armatura, che muove all'assalto, – era racchiuso uno dei capitoli più fitti d'avvenimenti della nostra storia. Non ci restava che azzardare delle congetture.[17]

In the course of the narrative, alternative possibilities of interpretation need to be constantly abandoned. The text thus includes a great number of crossroads indicating points where the narrative could have taken entirely different directions. To all these potential stories the mute presence of the other onlookers and potential narrators is added, each of them mutely testifying to the many different potential stories that might have been read from the cards. Finally, the narrator of our text is also a reader, and his activity says just as much about the active role of the reader in creating meaning in a text as it does about narrating or creating a text. His role as reader runs parallel to that of the actual readers of the text, to us, invited to create our own versions of the stories. This becomes evident through the fact that the narrator-interpreter's interpretations of the mute narratives are accompanied by the images of the tarot cards in the margins of the text. As a result, the reader may examine the interpretation the narrator offers with

---

16 *The Castle of Crossed Destinies*, trans. William Weaver (New York and London: Harcourt Brace Jovanovich, 1976), 11-12. Translator's italics. My explanation. All quotations are taken from this edition, henceforth cited as *Castle*.

17 *Il castello dei destini incrociati* (Torino: Einaudi 1973), 11-12. Italics I. C. All quotations in Italian are taken from this edition, henceforth cited as *Castello*.

reference to the cards and develop his or her own version of the stories. The narrative in its present form is thus set against the many alternative narratives which could have been produced by different subjects with the same material, but which are missing. The mute onlookers function as reminders of the possibility of different perspectives. In this way, the narrative constantly implies a potential beyond itself; what it says points to what it leaves unsaid.

This idea is most graphically conveyed through the metaphor of the network or net which Calvino used in his definition of the novel: It is not only the strings and knots which characterize the net, but also the lacunae they circumscribe. The image of the net thus not only connotes relations and complexity, but also abstraction, omission and loss. We should always consider it, in the first place, against the background of the material world and the infinitely many relations that may be established among its elements, but also in its fullness of being, which no net, however finely woven, can imitate. The mute tarot cards – and, by implication, any text – represent a world in miniature, of which infinitely many different readings are possible. At the same time, they also represent a choice by the author. The text of the *Castle* stands thus both for a reading as well as for a world.

The interpretations of the rows of cards are characterized by limitations, and this not just because of the decisions that any narrative must inevitably make, but because each narrative derives from the solipsistic world of its narrator. Narrating, as it is characterized by the exchange of laying cards and interpreting them, constitutes not so much a method of communication as an act of introspection, the reader's recognition of her or his own self reflected in the text. This is made evident by the fact that, with the exception of our narrator-interpreter, the first concern of each participant lies in assembling his or her own story from the cards. Accordingly, the spectators' attention often lags; their main interest in the stories told by the others seems to be that of finding the most suitable spot to insert their own narrative, i.e., the spot that will allow them to reuse the cards that best express what they have to say.[18]

The solipsistic nature of both interpretation and narrative creations is also made manifest by the fact that the kind of stories our narrator extracts from the cards – that is, the only stories actually spelled out – either are very generic or are stories derived from the interpreter's memory. Thus, he *recognizes* the story of Orlando's madness and that of Astolfo's lunar journey in two of the mute narratives.[19] The latter story is, moreover, simply evoked by the desire he feels after 'reading' Orlando's story "to collect further testimony ... [o]n the subject of Roland's sanity, especially from the one who made his recovery a duty, a test for his own ingenious daring. I would have liked

[18] See the situation after the alchemist has told his story: "In any case, some of us allowed our minds to wander, or we lingered over certain couplings of cards and were unable to go on.

One of us, for example, a warrior with melancholy eyes, began toying with a *Page of Swords*, which strongly resembled him, and a *Six of Clubs*; and he placed them beside the *Seven of Coins* and *The Star*, as if he wanted to build up a vertical row on his own." (*Castle*, 21, translator's italics)

[19] See the narrator's comment on identifying Orlando: "We had recognized him;" and later: "we recognized Angelica". (*Ibid.*, 30)

him, Astolfo, to be there with us."[20] His narrative thus confirms Marco Polo's observation in *Invisible Cities*: "I speak and speak, ... but the listener retains only the words he is expecting. ... It is not the voice that commands the story: it is the ear."[21] This means that the meaning a reader recovers from a text is already part of his system, or at least that what determines it are the expectations the audience brings to a text. It is not entirely new. From the author's perspective, it means that her or his literary creations are determined by old memories and experiences, that they are basically a recycling of old material.

Nevertheless, the act of narrating (both as authors and readers) constitutes a fundamental human need in as much as it helps us to understand the world (both the real world and the world of the text) by imagining it in terms of a rational order. It allows us to place objects and facts into a relation with each other and, most importantly, with ourselves. It helps us to find ourselves in the world. Conversely, "the forest is self-loss, shuffling [*mescolanza*]."[22] The forest, where the characters have lost their faculty of speech, quite obviously symbolizes the chaotic, nonsensical world outside of any narrative order. As this statement makes clear, the chaos of the forest is also reflected in the shuffled cards before the narrators have arranged them in a meaningful order.

The loss of self associated with the chaotic world (the forest and the shuffled cards) and the necessity to restore order in this world through narrative are best illustrated by the fact that the mad Orlando is placed right at the center of the square of cards. This choice has an important metanarrative significance and needs to be read in the context of Orlando's original story: Ariosto also placed the madness of his protagonist right at the center of his work,[23] a work which, because of the complex multiplicity of the narrative threads combined in it, he refers to as "la grande *tela* ch'io lavoro".[24] The metaphor of the net is significant in our context, especially if we read it in relation to the parallelism that the narrator of the *Furioso* establishes right from the beginning of his work between Orlando's *furore*, or madness, and his own. The narrator's madness has two aspects: On the one hand, there is his amorous frenzy, a milder edition of the total madness caused by Orlando's wretched love for Angelica; but then his madness is also a *furor poeticus* and denotes the author's creative inspiration, from which the work originates.

This second aspect of the narrator's madness also relates to Orlando's madness, as I would like to show in my reading of the central chapter of the poem, which focuses on the moment Orlando turns mad. The narrator of the *Furioso* here persistently refers to the pieces of writing that Orlando comes across wherever he goes, texts with which Angelica and her lover Medoro have adorned trees, bushes and rocks – as if to say that

20 *Ibid.*, 35. Adaptation mine. – "Sul senno di Orlando mi sarebbe piaciuto raccogliere altre testimonianze, soprattutto da colui che del recupero s'era fatto un dovere, una prova per il suo ardire ingegnoso. Avrei voluto che fosse lí con noi Astolfo." (Castello, 35)

21 *Cities*, 135. – "Io parlo, parlo, ... ma chi m'ascolta ritiene solo le parole che aspetta. ... Chi comanda al racconto non è la voce: è l'orecchio." (*Città*, 143)

22 "Il bosco è la perdita di sé, mescolanza." – I have given my own translation because Weaver translates *mescolanza* with 'mingling' rather than 'shuffling'; but it is vital that the word *mescolanza* is also used with reference to cards, whereas 'mingling' is not. (*Castello*, 13)

23 Orlando turns mad in Canto 23 of the 46 cantos the work contains.

24 'the great *fabric* that I am creating'. (*Orlando furioso*, xiii, 81, 2, emphasis mine)

Orlando's madness was caused by these very texts and not just by what they give proof of: Angelica's love for another man. The parallelism between the texts by the two lovers and the *'carte'* the poet keeps filling with his verses is emphasized even more by the Arabic verses scratched into the rock at the entrance of the grotto and the shepherd's cabin where the two lovers were engaged in amorous play and where Orlando "each wall, each door, each window finds filled with the hateful writing."[25] This is a powerful illustration of the poet's own situation at this particular point in his composition: Having arrived at the chaotic center of his net, he needs to work the many threads involved in his complex plot into a completed whole. Analogically, the Orlando character in the *Castle*, having arrived at "the chaotic heart of things, the center of the square of the cards and of the world, the point of intersection of all possible orders",[26] needs to connect his narrative to the ring of cards – the en*cyclo*pedia! – that surrounds him, and to leave the forest behind him in full consciousness.[27]

It is not by coincidence, however, that the most beautiful and suggestive tarot cards of all (Orlando kept them all for himself) have been placed at the center of the square. The reduction of the tangled connections and meanings to a narrative thread also brings disenchantment. At the border of the square of cards, Orlando is awaited by *Reason*:

> Was that blond dispenser of justice with her sword and scales the image of Reason, with whom in any event he would finally have to settle his accounts [*finire per fare i conti*]? Was she the Reason of the story, lurking under the combining Chance of the scattered tarots? Did this mean that, however he may wander, the moment comes when they catch him, Roland, and bind him, and force down his throat the intellect he has rejected?[28]

> Era dunque l'immagine della Ragione quella bionda giustiziera con spada e bilancia con cui lui doveva in ogni caso finire *per fare i conti*? Era la Ragione del racconto che cova sotto il Caso combinatorio dei tarocchi sparpagliati? Voleva dire che comunque giri poi viene il momento che lo acchiappano e lo legano, Orlando, e gli ricacciano in gola l'intelletto rifiutato?[29]

The double meaning of the phrase *fare i conti*, which not only means 'getting even with someone', but also '*making stories*', is vital.[30] However, as soon as the work is

---

25 "de l'odiato scritto ogni parete, ogni uscio, ogni finestra vede piena". (*Orlando furioso*, xxiii, 117, 3-4, translation mine)

26 *Castle*, 33. – "nel cuore caotico delle cose, al centro del quadrato dei tarocchi e del mondo, al punto d'intersezione di tutti gli ordini possibili." (*Castello*, 33)

27 Moreover, the narrator of the book is taken by the same maniacal (or poetic) frenzy to bring this work to its completion, whatever the losses – even if it means losing himself in all the other stories: "What is left me is only the manic determination to complete, to conclude, to make the sums work out [*a far tornare i conti*]. I still have to cover two sides of the quadrangle in the opposite direction, and I advance only out of punctilio, so as not to leave things half-done." (*Castle*, 46, my addition)

28 *Ibid.*, 33-34. Addition mine.

29 *Castello*, 34. Emphasis mine.

30 So, too, is the double meaning of the expression *fare tornare i conti* in the passage quoted in note 27. Along with its meaning of 'solving the problem', it describes both the narrative structure of the

completed and each card woven into the net of stories, the strings of the net dissolve and the stories become indistinguishable. The order of relations set up among the cards by means of the narratives disappears, and the square of cards turns back into a chaotic forest. In this context, Faust's comment in the *Tavern of Crossed Destinies*, a text written according to the same principle as the *Castle* and included in the same book, is instructive:

> The world does not exist, ... there is not an all, given all at once: there is a finite number of elements whose combinations are multiplied to billions of billions, and only a few of these find a form and a meaning and make their presence felt amid a meaningless, shapeless dust cloud [*un pulviscolo senza senso e senza forma*]; like the seventy-eight cards of the tarot deck in whose juxtapositions sequences of stories appear and are then immediately undone.[31]

> – Il mondo non esiste, ... non c'è un tutto dato tutto in una volta: c'è un numero finito d'elementi le cui combinazioni si moltiplicano a miliardi di miliardi, e di queste solo poche trovano una forma e un senso e s'impongono in mezzo a un pulviscolo senza senso e senza forma; come le settantotto carte del mazzo di tarocchi nei cui accostamenti appaiono sequenze di storie che subito si disfano.[32]

In referring to the 'world', Faust means a complete and ordered world of narrative. By contrast, the real world is '*un pulviscolo senza senso e senza forma*', symbolized by the image of the crystal of tarot cards turned opaque. To keep up the order of relations and keep it distinguished from the confusing world of images and essence, the stories would have to be constantly retold. This is only possible in the artificial setting of a contained and clearly circumscribed space, in the closed encyclopedia of the tarot deck, which sets the frame for Calvino's text. At the same time, this encyclopedia is also broken open, leaving a passage to the world outside, as is clearly symbolized by the disconnected drawbridge leading into the castle ("*ponte levatoio sconnesso*").[33]

Calvino's text, it is true, is an entirely artificial construct, governed by strict conventions; it is a closed world of limited possibilities. And yet by focusing on its own conventionality, on its choices and omissions, it makes room for alternatives, for divergent points of view, and leaves an open passage to the world unordered by narrative. Let me now turn to Calvino's bigger and more complex crystal, the *Invisible Cities*, and see how this text defines its encyclopedic potential.

Powerful and isolated, the great emperor Kublai Khan is often overcome by a terrible sense of emptiness when he tries to imagine his empire. In spite of the many reports he receives from the many explorers and ambassadors from the different provinces under his rule, his empire has reached dimensions that leave Kublai only with the

---

mute narratives with their double rows of cards, where the turning point and climax of the narrative coincides with the turning point between the two rows. It also describes the practice of inverting each narrative and telling it again by beginning at the tail, its palindromic reading, so to speak.

31 *Castle*, 97. Addition mine.

32 *Castello*, 97.

33 *Castello,* 3. Weaver imprecisely translates "rattling drawbridge". (*Castle*, 3)

impression of "an endless, formless ruin".[34] He seeks consolation in his conversations with his favorite ambassador, the Venetian traveler Marco Polo, hoping that from this man's idiosyncratic accounts he may recover a strategy to comprehend his empire in its totality. In one of his euphoric moments, the great Khan insists that his empire is like a perfect crystal:

> 'And yet I know,' he would say, 'that my empire is made of the stuff of crystals, its molecules arranged in a perfect pattern. Amid the surge of the elements, a splendid hard diamond takes shape, an immense, faceted, transparent mountain.'"[35]

> – Eppure io so, – diceva, – che il mio impero è fatto della materia dei cristalli, e aggrega le sue molecole secondo un disegno perfetto. In mezzo al ribollire degli elementi prende forma un diamante splendido e durissimo, un'immensa montagna sfaccettata e trasparente.[36]

And he reproaches Marco for hiding from him the greatness of his destiny by describing to him only "disappointing appearances" and "inessential melancholies".[37] As always, the eloquent Venetian is not at a loss for an answer; yet rather than look at the specific answer he gives Kublai at this point in the text, I suggest we look for the answer in the book at large.

On the surface, the *Invisible Cities* itself seems like a crystal: The book is subdivided into nine parts each of which contains five descriptions of a city, with the exception of the first and the last chapter, where ten cities are described. Every chapter is framed by two short texts printed in italics presenting extracts from conversations between Kublai and Marco Polo. In these passages, the two interlocutors speak about the cities of the empire as well as methods of describing them. Each city bears a woman's proper name, while it shares its title and a number assigned to it with some of the other cities. These numbers and titles are neatly arranged in the 'table of contents' that precedes the text.

The neat organization of the book and the categorization of the many cities into different themes such as 'cities and memory', 'cities and desire', 'cities and signs', 'subtle cities', etc., seem to offer help to both Khan and the reader (for the two are in many respects synonymous)[38] so that they may find their way through a maze of cities. Yet the crystalline structure is deceptive. Marco's descriptions, rather than furnishing Kublai with the desired comprehensive representation of his empire, convey the idea of

---

34 *Ibid.*, 13. – "uno sfacelo senza fine né forma". (*Castello*,13)

35 *Cities*, 60.

36 *Città*, 66.

37 *Cities*, 60. – "delusive apparenze", "malinconie inessenziali". (*Città*, 66)

38 The parallelism between Kublai and the reader is made evident on the very first page, in the passage describing the sense of emptiness that follows the satisfaction emperors feel after great conquests. The reference to the emperors suddenly turns to a collective 'we': "*In the lives of emperors there is a moment which follows pride in the boundless extension of the territories we have conquered, and the melancholy and relief of knowing we shall soon give up any thought of knowing and understanding them. There is a sense of emptiness that comes over us at evening, with the odor of the elephants after the rain and the sandalwood ashes growing cold in the braziers*". (*Cities*, 5)

the impossibility of such an enterprise. Thus, Marco repeatedly starts his description of a city by saying how he might describe this city without saying anything about its true character. He then goes on to say what its accurate description would have to include, yet such a description would clearly be impossible to give. A good example is the city of Zaira, about which Marco says:

> In vain, great-hearted Kublai, shall I attempt to describe Zaira, city of high bastions. I could tell you how many steps make up the streets rising like stairways, and the degree of the arcades' curves, and what kind of zinc scales cover the roofs; but I already know this would be the same as telling you nothing. The city does not consist of this, but of the relationships between the measurements of its space and the events of its past ... A description of Zaira as it is today should contain all Zaira's past. The city, however, does not tell its past, but contains it like the lines of the hand, written in the corners of the streets, the gratings of the windows, the banisters of the steps, ...[39]

> Inutilmente, magnanimo Kublai, tenterò di descriverti la città di Zaira. Potrei dirti di quanti gradini sono le vie fatte a scale, di che sesto gli archi dei porticati, di quali lamine di zinco sono ricoperti i tetti; ma so già che sarebbe come non dirti nulla. Non di questo è fatta la città, ma di relazioni tra le misure del suo spazio e gli avvenimenti del suo passato ... Una descrizione di Zaira quale è oggi dovrebbe contenere tutto il passato di Zaira. Ma la città non dice il suo passato, lo contiene come le linee d'una mano, scritto negli spigoli delle vie, nelle griglie delle finestre, negli scorrimano delle scale ...[40]

Of the city called Dorotea, Marco says that there are two ways of speaking about it: to give a generic description of its basic structure and some old customs practiced among its inhabitants; or to describe the very personal impressions a stranger had of Dorotea when he arrived there for the first time, such as those of the cameleer, who allegedly told Marco:

> I arrived here in my first youth, one morning, many people were hurrying along the streets toward the market, the women had fine teeth and looked you straight in the eye, three soldiers on a platform played the trumpet, and all around wheels turned and colored banners fluttered in the wind. Before then I had known only the desert and the caravan routes. In the years that followed, my eyes returned to contemplate the desert expanses and the caravan routes; but now I know this path is only one of the many that opened before me on that morning in Dorothea.[41]

> Vi arrivai nella prima giovinezza, una mattina, molta gente andava svelta per le vie verso il mercato, le donne avevano bei denti e guardavano dritto negli occhi, tre soldati sopra un palco suonavano il clarino, dappertutto intorno giravano ruote e sventolavano scritte colorate. Prima d'allora non avevo conosciuto che il deserto e le piste delle carovane. Quella mattina a Dorotea sentii che non c'era bene della vita che non potessi aspettarmi. Nel seguito degli anni i miei occhi sono tornati a

---

[39] *Cities*, 10-11.
[40] *Città*, 18.
[41] *Cities*, 9.

> contemplare le distese del deserto e le piste delle carovane; ma ora so che questa è solo una delle tante vie che mi si aprivano quella mattina a Dorotea.[42]

Similarly, to the cameleer reaching the city of Despina from the desert and to the sailor reaching it from the sea, Despina appears in ways diametrically opposed to each other: While the cameleer imagines it like a ship that will carry him away from the desert, the sailor imagines it like a huge camel loaded with the most wonderful delicacies of the Orient.[43]

From the accounts of these three cities, we start to realize that a full representation of any city would have to include not just a description of its physical appearance in all its details and relations, but also an account of its past to show how it came to be the way it is perceived in the present. It would have to include the various perceptions of the city by each person who ever entered it, as well as the desires each stranger projected into the city, the possibilities of lives imagined in the city, and their nostalgic memories. It would have to include a description of everything that took place in the city and how it is, was and will be remembered. It would have to include the future of the city, imaginable, possible or actual. To have a full conception of the city in one's mind would entail envisioning all these elements in their totality and in their relations to each other. In short, it would require of a single mind to encompass the collective perceptions of the city as perceived by everyone who ever imagined it.

And all this would be required just to imagine a single city in its totality. For Kublai to have an adequate conception of his whole empire would then require the infinitely more gigantic task of having such an idea of every single city in his empire and of their relations among each other. This is obviously impossible. Still more impossible is Marco's task of representing such a conception to Kublai in their conversations; or that of the book of conveying it to the reader. Kublai's hopeless desire to imagine the totality of his empire in his mind symbolizes the author's hopeless desire to represent the world in the infinite complexity of its relations and the infinitely diverse possibilities of conceiving them, as well as his desire to accomplish a representation of the world as it is in its essence, beyond anyone's conception of it. The different cities in Kublai's empire each represent a full world in themselves; as a result, the idea of conceiving and representing their combined totality becomes all the more hopeless.

Looked at more closely then, *Invisible Cities* turns out to be a many-facetted collection of fragments which both complement and contradict each other. The text is crystal-like for the way in which the few things that are said about each city refract in an infinite spectrum of what could also be said, but is missing. It refracts into a potential multiplicity that remains implied even while it is hinted at. At the same time, the text is decidedly not transparent; it eludes *comprehension*. It is itself 'an endless, formless ruin' that defies comprehensive unity. The narrative does not convey any sense of progress, neither in the conversations between Kublai and Marco, nor in the latter's accounts of cities. We get the impression that, no matter how many more cities he adds, Marco will never bring his task of describing the Khan's empire any closer to completion.

---

42 *Città*, 17.

43 *Cities*, 17.

His descriptions of cities constantly communicate their essential elusiveness. The cities are "constant metonymies of desire". This is indicated by the feminine names given to all of the cities;[44] moreover, in some of Marco's accounts, desire and the lack or withdrawal it implies figure as the fundamental characteristic of the city in question. In the case of Zobeide, for instance, desire and withdrawal are written into the very structure of the city, thus becoming its text. The city was built by men from different nations who all had the same dream:

> They saw a woman running at night through an unknown city; she was seen from behind, with long hair, and she was naked. They dreamed of pursuing her. As they twisted and turned, each of them lost her. After the dream they set out in search of that city; they never found it, but they found one another; they decided to build a city like the one in the dream. In laying out the streets, each followed the course of his pursuit; at the spot where they had lost the fugitive's trail, they arranged spaces and walls differently from the dream, so she would be unable to escape again.
>
> This was the city of Zobeide, where they settled, waiting for that scene to be repeated one night. None of them, asleep or awake, ever saw the woman again.[45]

> videro una donna correre di notte per una città sconosciuta, da dietro, coi capelli lunghi, ed era nuda. Sognarono d'inseguirla. Gira gira ognuno la perdette. Dopo il sogno andarono cercando quella città; non la trovarono ma si trovarono tra loro; decisero di costruire una città come nel sogno. Nella disposizione delle strade ognuno rifece il percorso del suo inseguimento; nel punto in cui aveva perso le tracce della fuggitiva ordinò diversamente che nel sogno gli spazi e le mura in modo che non gli potesse piú scappare.
>
> Questa fu la città di Zobeide in cui si stabilirono aspettando che una notte si ripetesse quella scena. Nessuno di loro, né nel sonno né da sveglio, vide mai piú la donna.[46]

With a clever allusion to Ariosto's lunar episode, Zobeide is appropriately called "the white city, well exposed to the moon."[47] Similarly, in describing the city of Diomira, Marco first refers to the material beauties that the traveler is already familiar with from other cities. He then points out the essential characteristic of Diomira:

> But the special quality [*la proprietà*] of this city for the man who arrives there on a September evening, when the days are growing shorter and the multicolored lamps are lighted all at once at the doors of the food stalls and from a terrace a woman's voice cries ooh!, is that he feels envy toward those who now believe they have

---

44 Gian-Paolo Biasin, "Italo Calvino in Mexico: Food and Lovers, Tourists and Cannibals," in *PMLA* 108, no. 1 (1993): 77.

45 *Cities*, 45. Teresa de Lauretis refers to the description of Zobeide at the beginning of *Alice Doesn't: Feminism, Semiotics and Cinema* (Bloomington: Indiana University Press, 1984). She sees Calvino's text as "an accurate representation of the paradoxical status of women in Western discourse: while culture originates from woman and is founded on the dream of her captivity, women are all but absent from history and cultural process." (12-14) The motif of 'women at windows' which appears in many of Marco's accounts of cities could be said to be formulaic for the theme of desire that characterizes many of the cities described.

46 *Città*, 51.

47 *Cities*, 45. – "città bianca, ben esposta alla luna." (*Città*, 51)

once before lived an evening identical to this and who think they were happy, that time.[48]

Ma la *proprietà* di questa è che chi vi arriva una sera di settembre, quando le giornate s'accorciano e le lampade multicolori s'accendono tutte insieme sulle porte delle friggitorie, e da una terrazza una voce di donna grida: uh!, gli viene da invidiare quelli che ora pensano d'aver già vissuto una sera uguale a questa e d'esser stati quella volta felici.[49]

Not only does Marco express what he feels to be characteristic of the city with the fleeting spirit of a moment in the personal experience of an individual; this experience is itself characterized by absence and lack: The cry of satisfaction (or projected satisfaction) experienced by the desired woman (signifying lack) evokes the envy (signifying lack) of nostalgic memories of past happiness (again signifying lack) experienced by others (again a projection). What is ironically and paradoxically called the *proprietà* of Diomira (meaning not just 'characteristic', but also 'material possession') is thus the very epitome of lack.

The theme of absence and lack that these cities circumscribe manifests itself in Khan's inability to have an image of his empire in his mind, the 'sense of emptiness' he experiences whenever he tries to imagine this empire.[50] It also manifests itself in the frustrating sense of incomprehensibility we experience as readers whenever we try to join the various parts of the text into a comprehensive whole.[51] Calvino himself has described *Invisible Cities* as "a many-faceted structure in which each brief text is close to the others in a series that does not imply logical sequence or a hierarchy, but a network in which one can follow multiple routes and draw multiple ramified conclusions."[52] While the text thus invites the reader to look for ever new ways of traveling through it by drawing connections between its different elements, it defies final comprehension, any single interpretive logic. It refuses to make sense, but asks to be left and further developed in its multiplicity. What it proffers is constant denial.[53]

48 *Cities*, 7.

49 *Città*, 15. Emphasis mine.

50 *Cities*, 5.

51 In his review of the book, Guido Almansi called it "Le città illeggibili", thus triggering a debate about the book's illegibility. Cf. Christoph Hoch, "Eudossias Teppich: zu Theorie und Praxis intertextuellen Erzählens in Italo Calvinos *Le città invisibili*," in *Romanische Forschungen* 106 (1994): 189. Carol P. James has observed: "And as the Emperor Kublai Khan links his political power to his possession of the cities as stories, the reader takes pleasure in attempts to subjugate the stories by understanding their (possible) allegorical content." Cf. "The Fragmentation of Allegory in Calvino's *Invisible Cities*," in *The Review of Contemporary Fiction* 6, no. 2 (1986): 88.

52 *Memos*, 71.

53 Another text in which this kind of denial is programmatic is Melville's "Bartleby, the Scrivener", where Bartleby's friendly refusal to comply with any of the requests made by his employer and 'reader', his stubborn "*I would prefer not to*" (itself an incomplete sentence), is what keeps the text alive. The alternatives are brick walls (Wall Street) and death. In writing this chapter, my thoughts keep returning to a seminar offered by Hans-Jost Frey at the University of Zurich a couple a years ago. The seminar was on 'the secret in literature' and "Bartleby" was one of the texts discussed.

This becomes evident from the frequent discrepancies that characterize Marco's accounts of cities, both within themselves and between each other. Along with the discrepancies in the discourse about a city, there are those between the discourse and what it refers to, i.e., the city as it is in reality: As Marco once explicitly reminds Kublai, the concept of a city, however exhaustively annotated or diverse, must not be confused with the city itself, even though there is a relation between the two.[54] In referring to the city of Anastasia, Marco explains that no description of the many pleasures Anastasia offers will ever make it possible to express the "real essence" of the city, which again paradoxically turns out to be the 'fullness of desire', that is, the 'fullness of lack':

> for while the description of Anastasia awakens desires one at a time only to force you to stifle them, when you are in the heart of Anastasia one morning your desires waken all at once and surround you. The city appears to you as a whole where no desire is lost and of which you are a part, and since it enjoys everything you do not enjoy, you can do nothing but inhabit this desire and be content.[55]

> perché mentre la descrizione di Anastasia non fa che risvegliare i desideri uno per volta per obbligarti a soffocarli, a chi si trova un mattino in mezzo ad Anastasia i desideri si risvegliano tutti insieme e ti circondano. La città ti appare come un tutto in cui nessun desiderio va perduto e di cui tu fai parte, e poiché essa gode tutto quello che tu non godi, a te non resta che abitare questo desiderio ed esserne contento.[56]

The problem referred to here is, of course, the classic problem of all narrative: its linearity, which in *Invisible Cities* Calvino tries to break up as much as possible by placing the many short texts that make up his book in a random order. Yet this is not the only problem with the description of the city, for, as we learn at the end of the short text, what is called the 'real' experience of Anastasia, because everybody shares it, is an illusion: "Such is the power ... that Anastasia, the treacherous city, possesses; if for eight hours a day you work as a cutter of agate, onyx, chrysoprase, your labor which gives form to desire takes from desire its form, and you believe you are enjoying Anastasia wholly when you are only its slave."[57]

In the case of Tamara, the traveler's perception of the city is also distracted from what the city really is: Walking about Tamara, the traveler does not see things, but figures of things signifying other things. "If a building has no signboard or figure, its very form and the position it occupies in the city's order suffice to indicate its function: the palace, the prison, the mint, the Pythagorean school, the brothel. ... Your gaze scans

---

Reading Calvino's texts made me remember Frey's comment: "Maybe every text says a little bit 'I would prefer not to'."

54 *Cities*, 61.

55 *Ibid.*, 12.

56 *Città*, 20.

57 "Tale potere ... ha Anastasia, città ingannatrice: se per otto ore al giorno tu lavori come tagliatore d'agate onici crisopazi, la tua fatica che dà forma al desiderio prende dal desiderio la sua forma, e tu credi di godere per tutta Anastasia mentre non ne sei che lo schiavo." (*Ibid.*)

the streets as if they were written pages."[58] While the city, as a fundamentally human creation, may look to the traveler like the embodiment of a meaningful structure, organized according to an obvious logic, his 'reading' does not tell the traveler anything about the city itself: "However the city may really be, beneath this thick coating of signs, whatever it may contain or conceal, you leave Tamara without having discovered it."[59]

In addition to its superficiality, this semiotic system reveals its arbitrariness when the traveler arrives in the city of Ipazia: Though nothing in this city appears strange to him and he believes he recognizes all the familiar elements of a city, everything has a radically different meaning from the one he takes for granted. "Signs form a language, but not the one you think you know. I realized I had to free myself from the images which in the past had announced to me the things I sought: only then would I succeed in understanding the language of Hypatia."[60] The alien logic of Ipazia recalls Jorge Luis Borges' taxonomy of animals, which Michel Foucault quotes in his Preface to *The Order of Things*: The classificatory system of animals allegedly quoted from a 'certain Chinese encyclopedia' appears to us as some kind of monstrous disorder. "In the wonderment of this taxonomy," says Foucault, "the thing we apprehend in one great leap, the thing that, by means of the fable, is demonstrated as the exotic charm of another system of thought, is the limitation of our own."[61] This is just what the traveler comes to realize in Ipazia.

Finally, when the traveler arrives in the city of Zoe, he is not just disappointed in his expectation of recognizing a basic order common to all cities – Zoe does not show any system of organization whatsoever. All its elements are indistinguishable from all the others; nothing has a purpose of its own. The city is unthinkable and the traveler cannot help mixing up its points in his mind. "But why, then, the city? What *line separates* the inside from the outside, the rumble of wheels from the howl of wolves?"[62] Here, the city becomes the metaphor for the mental framework by which it is conceived and outside of which it loses its right to exist. It becomes its own metaphor. Yet only in the mind of the traveler who tries to comprehend it: the 'line that separates' may be meaningful, but it is not *essential*. And the city's justification remains an issue that cannot negate the city's manifest, if elusive existence.

Throughout his conversations with Kublai Khan, Marco keeps emphasizing, first of all, that the infinitely many and diverse ways of experiencing and perceiving a city, an empire, the world, are beyond description and comprehension and that, moreover, no mental concept of a city, however potentially complete, can comprehend its reality, its

---

58 *Cities*, 13. – "Se un edificio non porta nessuna insegna o figura, la sua stessa forma e il posto che occupa nell'ordine della città bastano a indicarne la funzione: la reggia, la prigione, la zecca, la scuola pitagorica, il bordello. ... Lo sguardo percorre le vie come pagine scritte." (*Città*, 22)

59 *Cities*, 14. – "Come veramente sia la città sotto questo fitto involucro di segni, cosa contenga o nasconda, l'uomo esce da Tamara senza averlo saputo." (*Città*, 23)

60 *Cities*, 48. – "I segni formano una lingua, ma non quella che credi di conoscere. – Capii che dovevo liberarmi dalle immagini che fin qui m'avevano annunciato cose che cercavo: solo allora sarei riuscito a intendere il linguaggio di Ipazia." (*Città*, 53-54)

61 *The Order of Things: An Archaeology of the Human Sciences* (New York: Vintage Books, 1994), xv.

62 *Cities*, 34. – "Ma perché allora la città? Quale *linea separa* il dentro dal fuori, il rombo delle ruote dall'ululo dei lupi?" (*Città*, 40, emphasis mine)

brute material facticity: Just as one's own idea of a city can never comprehend the innumerable other possible ideas of that city, so must the idea of a city be distinguished from its reality. To have an idea of a city in one's mind does not mean that this idea equals the reality of the city or is its copy, nor does it mean that, as a concept, this idea is of universal validity. It is just a personal fiction. It has nothing to do with *possessing* the city, neither mentally nor physically. This is what Marco's narratives keep saying as well as implying throughout.[63] Accordingly, the portraits of cities Marco offers to Kublai are mental exercises with which Marco sketches an imaginary empire as diverse in its aspects as he himself is capable of imagining. The obvious limitations of this sketch could be said to form part of the program. Indeed, although the emperor suspects his 'ambassador' of offering him mere permutations of the same elements,[64] and cannot help having the impression that Marco is speaking primarily about Venice, his native city, this does not seem to bother Kublai all that much, and it clearly does not inconvenience the two of them in their activity. Indeed, the more obvious the limitations, the more evident and pressing the task of leaving room for what is left out and implying it as a potential. As it turns out, this is achieved most successfully when the means of communication between Marco and Kublai are most limited:

> *Newly arrived and totally ignorant of the Levantine language[s], Marco Polo could express himself only with gestures, leaps, cries of wonder and of horror, animal barkings and hootings, or with objects he took from his knapsacks – ostrich plumes, pea-shooters, quartzes – which he arranged in front of him like chessmen. ... the ingenious foreigner improvised pantomimes that the sovereign had to interpret: one city was depicted by the leap of a fish escaping the cormorant's beak to fall into a net; another city by a naked man running through fire unscorched; a third by a skull, its teeth green with mold, clenching a round, white pearl. The Great Khan deciphered the signs, but the connection between them and the places visited remained uncertain; he never knew whether Marco wished to enact an adventure that had befallen him on his journey, an exploit of the city's founder, the prophecy of an astrologer, a rebus or a charade to indicate a name.*[65]

> *Nuovo arrivato e affatto ignaro delle lingue del Levante, Marco Polo non poteva esprimersi altrimenti che con gesti, salti, grida di meraviglia e d'orrore, latrati o chiurli d'animali, o con oggetti che andava estraendo dalle sue bisacce: piume di struzzo, cerbottane, quarzi, e disponendo davanti a sé come pezzi degli scacchi. ... l'ingegnoso straniero improvvisava pantomime che il sovrano doveva interpretare: una città era designata dal salto d'un pesce che sfuggiva al becco del cormorano per cadere in una rete, un'altra città da un uomo nudo che attraver-*

---

63 Similarly, in *Palomar*, Calvino's last novel, the readability of the world proves to be an illusion and Palomar's unflagging attempts to decipher the universe are constantly frustrated. Cf. Christine Lessle, *Weltreflexion und Weltlektüre in Italo Calvinos erzählerischem Spätwerk* (Bonn: Romanischer Verlag, 1992), 167-70.

64 See "Kublai had noticed that Marco Polo's cities resembled one another, as if the passage from one to another involved not a journey but a change of elements". (*Cities*, 43) See also the city of Clarice, of which "[o]nly this is known for sure: a given number of objects is shifted within a given space, at times submerged by a quantity of new objects, at times worn out and not replaced; the rule is to shuffle them each time, then try to assemble them." (*Ibid.*, 108, adaptation mine)

65 *Ibid.*, 21-22. Correction mine.

> *sava il fuoco senza bruciarsi, una terza da un teschio che stringeva tra i denti verdi di muffa una perla candida e rotonda. Il Gran Kan decifrava i segni, però il nesso tra questi e i luoghi visitati rimaneva incerto: non sapeva mai se Marco volesse rappresentare un'avventura occorsagli in viaggio, una impresa del fondatore della città, la profezia d'un astrologo, un rebus o una sciarada per indicare un nome.*[66]

Marco's ingenious performances have the power of emblems in so far as they cannot be forgotten, but since they remain ambiguous, they may change and develop their meanings. Just like the tarot cards in the mute accounts of the travelers in the *Castle*, they offer receptacles for endless new meanings and thus represent potentially inexhaustible sources of meaning. Yet the most important point is that these receptacles are *empty* to start with and that every meaning attributed to them derives from the mind of an interpreter. Therefore, when Kublai conceives a city based on one of Marco's emblematic performances, this city is the creation of his own imagination. To return once again to the image of the net with which Calvino defined the contemporary novel and which he used to describe the many different routes that can be traced through the collection of texts that the book presents: This net is again what is missing or what both Marco's emblematic performances and his fragmentary accounts of cities merely imply by representing points of departures, possible beginnings from which such a net may be realized in the imagination of the reader and Kublai. It is this aspect that makes Marco's performances most precious to Kublai:

> *But what enhanced for Kublai every event or piece of news reported by his inarticulate informer was* the space that remained around it, the void *not filled with words. The descriptions of cities Marco Polo visited had this virtue: you could wander through them in thought, become lost, stop and enjoy the cool air, or run off.*[67]

> *Ma ciò che rendeva prezioso a Kublai ogni fatto o notizia riferito dal suo inarticolato informatore era* lo spazio che restava loro intorno, un vuoto *non riempito di parole. Le descrizioni di città visitate da Marco Polo avevano questa dote: che ci si poteva girare in mezzo col pensiero, perdercisi, fermarsi a prendere il fresco, o scappare via di corsa.*[68]

Here, the empty space is liberating and evokes an experience quite different from the desperate feeling of emptiness that keeps overcoming Kublai whenever he seeks to possess a clearly defined picture of his empire in his mind. This emptiness is liberating because, rather than signifying lack, it harbors possibilities. It offers a space where the imagination can be creative and is not just at a loss and where, at the same time, there is always room left for alternatives beyond the limited perspective of any individual or a collective imagination; where there is room for things as they are, beyond imagination and interpretation.

---

66 *Città*, 29-30.
67 *Cities*, 38. Italics in the original. Emphasis mine.
68 *Città*, 45. Emphasis mine.

As time passes, Marco grows more and more proficient in the emperor's language and soon he is able to communicate verbally with the Khan. Yet their communication seems less felicitous than before, and very often words simply fail Marco. So he gradually returns to communicating by means of gestures, grimaces and looks. However, the great emperor's hunger for possession is not so easily appeased, and he keeps asking Marco for a method that will allow him to know his empire once and for all. Yet every time the unpredictable Venetian disappoints his aspirations: When Kublai asks him whether he will finally possess his empire once he knows the emblems for every city, Marco simply answers: "Sire, do not believe it. On that day you will be an emblem among emblems."[69] Soon, he comes up with another idea.

In front of Kublai's throne, there is a checkered majolica floor on which Marco displays the merchandise brought back from the empire's borders. From the way the Venetian arranges objects on the black and white tiles and changes their positions with studied moves, the monarch is reminded of figures on a chessboard: "*following Marco's movements, he observed that certain pieces implied or excluded the vicinity of other pieces and were shifted along certain lines.*"[70] By interpreting the different objects as chess pieces, Kublai starts to figure out possible orders of their arrangement. As a result, he comes to believe that the rules of chess imply the invisible order that stands at the basis of each city, the rules that govern its rise and fall, while the pieces offer symbols by means of which its various parts may be represented in all their potential diversity.[71] And he hopes with this metaphor to have found the key to possessing his empire at last:

> *If each city is like a game of chess, the day when I have learned the rules, I shall finally possess my empire, even if I shall never succeed in knowing all the cities it contains.*[72]
>
> *Se ogni città è come una partita di scacchi, il giorno in cui arriverò a conoscerne le regole possiederò finalmente il mio impero, anche se mai riuscirò a conoscere tutte le città che contiene.*[73]

Instead of sending Marco on expeditions to remote parts of his empire, he now engages him in endless games of chess:

---

[69] *Cities*, 23. – "Sire, non lo credere: quel giorno sarai tu stesso emblema tra gli emblemi." (*Città*, 30)

[70] *Cities*, 121. – "seguendo i gesti di Marco osservava che certi pezzi implicavano o escludevano la vicinanza d'altri pezzi e si spostavano secondo certe linee." (*Città*, 127)

[71] Christoph Hoch has pointed out that by comparing Marco's mute narratives with chess, Calvino is referring to Saussure's structural analysis of the system of language, where the rules of chess correspond to the grammar of a language (*langue*), while the individual moves and possible alternatives open to the players correspond to the concrete speech act (*parole*). Cf. "Eudossias Teppich," 205. The references to structural linguistics and semiotics are all too obvious throughout Calvino's text; yet I am much more interested in the way the arrangement of figures on a chess board and the rules of chess are used as mental frameworks for the representation of complex systems of relations.

[72] *Cities*, 121.

[73] *Città*, 127.

> *Knowledge of the empire was hidden in the pattern drawn by the angular shifts of the knights, by the diagonal passages opened by the bishop's incursions, by the lumbering, cautious tread of the king and the humble pawn, by the inexorable ups and downs of every game.*[74]

> *La conoscenza dell'impero era nascosta nel disegno tracciato dai salti spigolosi del cavallo, dai varchi diagonali che s'aprono alle incursioni dell'alfiere, dal passo strascicato e guardingo del re e dell'umile pedone, dalle alternative inesorabili d'ogni partita.*[75]

Yet if, with the 'inexorable alternatives' of each new game, chess offers an ideal metaphor to express the infinite multiplicity of phenomena and the possible correlation between them, its total appropriation and perfection as an abstract concept (represented symbolically in the checkmate) only brings back the sense of emptiness and loss that comes with conquests. Accordingly, Kublai realizes with resignation that the conclusion of each game brings nothing but the empty field under the feet of the conquered king. Abstraction, it seems, leads nowhere.

Once again, Marco helps the sovereign out of his depression and shows him how this seemingly empty field reveals its own characteristic features that mark its relations to the world. Marco first points out that the emperor's chessboard is an inlay of two woods, ebony and maple; from the wooden square Kublai's eyes are focused on, he then extracts the story of the tree from which this square originates. He concludes that it was cut from a stratum of the stem that had grown in a year of drought, then points to a knot, barely visible, where a bud had tried to shoot out, and to another mark that could have been the 'nest' of a larva: Judging from how small the mark is, it must have been the larva of a caterpillar that nibbled away at the leaves of the tree and was the reason why the tree was eventually felled.

> *The quantity of things that could be read in a little piece of smooth and empty wood overwhelmed Kublai; Polo was already talking about ebony forests, about rafts laden with logs that come down the rivers, of docks, of women at the windows...*[76]

> *La quantità di cose che si potevano leggere in un pezzetto di legno liscio e vuoto sommergeva Kublai; già Polo era venuto a parlare dei boschi d'ebano, delle zattere di tronchi che discendono i fiumi, degli approdi, delle donne alle finestre...*[77]

In this way, the little wooden square not only implies the history of the tree from which it was taken, but becomes the beginning of a potentially endless narrative involving the world. Marco's elegant *tour de force* shows how the most abstract of abstractions can always be saturated and brought to life again – indeed, not just that it *can*, but that it *must* be. Thus, his narrative of the field on the chessboard offers a practical translation of a crucial argument Marco keeps returning to throughout his conversations with

---

[74] *Cities*, 122-23.

[75] *Città*, 128.

[76] *Cities*, 132.

[77] *Città*, 140.

Kublai: that no abstract concept we construct in our imagination must ever be conceived as an end in itself. Rather, it should be allowed to point beyond itself, to remain connected with the endlessly complex tangle of worldly relations and the innumerable possible ways to conceive them. Moreover, it should leave room for a reality beyond the conception of any mind. We should regard any concept we create of something as a potential to be endlessly developed. A mental concept of something that we keep in the mind as a *possession* and treat as a perfect substitute of what we take it to represent is a dead letter.[78]

A good illustration of this argument is the city of Zora, which "has the quality [*la proprietà*!] of remaining in your memory point by point, in its succession of streets, of houses along the streets, and of doors and windows in the houses, though nothing in them possesses a special beauty or rarity."[79] As a result, the city that cannot be canceled from the mind itself becomes a mnemonic structure: "an armature, a honeycomb in whose cells each of us can place the things he wants to remember: names of famous men, virtues, numbers, vegetable and mineral classifications, dates of battles, constellations, parts of speech."[80] Yet when Marco sets out to visit the city, he cannot find it: "forced to remain motionless and always the same, in order to be more easily remembered, Zora has languished, disintegrated, disappeared. The earth has forgotten her."[81]

Of course, this is not to say that abstractions are negative and ought to be avoided. Obviously, the ability to abstract is a fundamental function of thinking. What it suggests is that in forming abstract concepts of facts, things or people, we must keep in mind that the concept we create in our mind is not a true representation, but merely a personal fiction that enables us to think and communicate. The abstract concept should always preserve a potential to be developed into an endless narrative that involves the world.

In his essay "Exactitude," Calvino speaks about two antithetical impulses that characterize his writing – his love for rationalization and abstraction and his love for intricate detail:

> my writing has always found itself facing two divergent paths that correspond to two different types of knowledge. One path goes into the mental space of bodiless rationality, where one may trace lines that converge, projections, abstract forms, vectors of force. The other path goes through a space crammed with objects and attempts to create a verbal equivalent of that space by filling the page with words, involving a most careful, painstaking effort to adapt what is written to what is not written, to the sum of what is sayable and not sayable. These are two different drives toward exactitude that will never attain complete fulfillment, one because

---

78 Poor Bartleby comes to mind again ...

79 *Cities*, 15. Addition mine. – "ha la *proprietà* di restare nella memoria punto per punto, nella successione delle vie, e delle case lungo le vie, e delle porte e delle finestre nelle case, pur non mostrando in esse bellezze o rarità particolari." (*Città*, 23, emphasis mine)

80 *Cities*, 15. – "un'armatura o reticolo nelle cui caselle ognuno può disporre le cose che vuole ricordare: nomi di uomini illustri, virtú, numeri, classificazioni vegetali e minerali, date di battaglie, costellazioni, parti del discorso." (*Città*, 23.)

81 *Cities*, 16. – "obbligata a restare immobile e uguale a se stessa per essere meglio ricordata, Zora languí, si disfece e scomparve. La Terra l'ha dimenticata." (*Città*, 24)

> 'natural' languages always say something *more* than formalized languages can – natural languages always involve a certain amount of noise that impinges upon the essentiality of the information – and the other because, in representing the density and continuity of the world around us, language is revealed as defective and fragmentary, always saying something *less* with respect to the sum of what can be experienced.[82]

In *Invisible Cities*, abstraction and concrete detail always stand side by side. This is shown in emblematic terms in the exchange between Marco and Kublai just referred to: Faced with Kublai's attempt to reduce his understanding of his empire to the combinatorics of chess pieces on a chessboard, Marco counters with the potentially endless story extracted from a single field on the board. Yet it is also shown in every single one of his accounts of cities, where the obviously fragmentary nature of the description, typically limited to sketching what an accurate description of the city in question *ought to* include, is contrasted with references to very specific images and sensations. This contrast between careful detail and omission is present on the very first page of the book, where the narrator refers to the terrible sense of emptiness that overcomes Kublai when he thinks of his empire: "*a sense of emptiness that comes over us at evening, with the odor of the elephants after the rain and the sandalwood ashes growing cold in the braziers.*"[83] The sense of emptiness is here contrasted with a description of a particular atmosphere involving a mixture of very specific smells.

To conclude this chapter, I would like to turn once more to a net, this time a tapestry claimed to represent a 'perfect' abstract design of reality: In the city of Eudossia a tapestry preserved by the inhabitants of the city represents the "true form of the city". Though at first sight nothing seems to resemble Eudossia less than the figures on this tapestry, its geometrical design shows all things contained in the city in their "true relations".

> Every inhabitant of Eudoxia compares the carpet's immobile order with his own image of the city, an anguish of his own, and each can find, concealed among the arabesques, an answer, the story of his life, the twists of fate.[84]

> Ogni abitante di Eudossia confronta all'ordine immobile del tappeto una sua immagine della città, una sua angoscia, e ognuno può trovare nascosta tra gli arabeschi una risposta, il racconto della sua vita, le svolte del destino.[85]

The abstract, unchanging design of this tapestry forms a text from which any personal narrative imaginable can be developed. Its completeness lies not so much in what it shows as in what the people looking at it read into its design. It contains an unimaginable multiplicity of narratives as its potential; its abstract design can always be constructed into 'a network of connections between the events, the people, and the things of the world'. Yet in spite of its apparent perfection in the eyes of each inhabi-

---

82 *Memos*, 74-75.

83 *Cities*, 5. – "*un senso come di vuoto che ci prende una sera con l'odore degli elefanti dopo la pioggia e della cenere di sandalo che si raffredda nei bracieri.*" (*Città*, 13)

84 *Cities*, 96-97.

85 *Città*, 103-4.

tant, the 'immobile order' of the tapestry turns out to be an unlikely, indeed deceitful representation of the city "as it is": "a stain that spreads out shapelessly, with crooked streets, houses that crumble one upon the other amid clouds of dust, fires, screams in the darkness."[86] It seems that the only way to do justice to the true reality of things is to keep drawing attention to the artificiality of representation, to its incompatibility with what is represented. The metaphorical image of the net thus offers a suitable image not just for a text, but for any mental concept of reality: It is through the holes and omissions in its structure, which remain, no matter how finely the net is wrought, that reality is allowed to shine as a mute presence.

---

[86] *Cities*, 97. – "una macchia che dilaga senza forma, con vie tutte a zigzag, case che franano una sull'altra nel polverone, incendi, urla nel buio." (*Città*, 104)

According to Christoph Hoch, Eudossia represents a kind of "microcosm" within the book as a whole in that its tapestry defines the *Invisible Cities* as an intertextual collage. For this argument he offers convincing evidence: Eudossia, he explains, is the name of an early Byzantine empress, the wife of Theodosios II, who made a name for herself in the history of literature with her Homeric centos. The name 'centos', which is derived from the Greek '*kentron*' or Latin '*cento*', meaning a rug made of patches, refers to patchwork poems composed from the verses of various other poets, mostly well-known ones such as Homer and Virgil. ("Eudossias Teppich," *passim* and especially 196)

Bibliography

Primary Sources:

Ariosto, Ludovico. *Orlando furioso*. Edited by Cesare Segre. 2 vols. Milan: Mondadori, 1976.

——. *Orlando furioso: Translated into English Heroical Verse by Sir John Harington*. (1591). Edited by Robert McNulty. Oxford: Clarendon, 1972.

Aristotle. *The Rhetoric and Poetics of Aristotle*. Edited by Friedrich Solmsen. New York: Random House, 1954.

Borel, Pierre. *Discours nouveau prouvant la pluralité des Mondes, que les astres sont des terres habitées, & la terre une Etoile, qu'elle est hors du centre du monde dans le troisiesme Ciel, & se tourne devant les soleil qui est fixe, & autres choses tres curieuses*. Geneva, 1657.

Browne, Sir Thomas. *The Major Works*. Edited by C. A. Patrides. Harmondsworth: Penguin, 1977.

Calvino, Italo. *Il castello dei destini incrociati*. Torino: Einaudi 1973.

——. *The Castle of Crossed Destinies*. Translated by William Weaver. New York and London: Harcourt Brace Jovanovich, 1976.

——. *Le città invisibili*. Torino: Einaudi, 1972.

——. *Invisible Cities*, Translated by William Weaver. New York and London: Harcourt Brace Jovanovich, 1972.

——. *Six Memos for the Next Millennium*. The Charles Eliot Norton Lectures 1985-86. Cambridge, MA: Harvard University Press, 1988.

Campanella, Tommaso. *La città del Sole*. In *Opere di Giordano Bruno e di Tommaso Campanella*. Edited by Augusto Guzzo and Romano Amerio. Milan and Napels, 1956.

Cebes, *Tabula of Cebes*. Greek and English edition. Edited and translated by John T. Fitzgerald and L. Michael White. Chico, CA: Scholars Press, 1983.

de Cyrano de Bergerac, Savinien. *Les oeuvres libertines de Cyrano de Bergerac*. (1921). Edited by Frédéric Lachèvre. Geneva: Slatkine Reprints, 1968.

——. *Other Worlds. The Comical History of the States and Empires of the Moon and the Sun*. Translated by Geoffrey Strachan. London: Oxford University Press, 1965.

Cicero, Marcus Tullius. *Somnium Scipionis*. In *De re publica*. Edited and translated by Clinton Walker Keyes. Loeb Classical Library. Cambridge, MA: Harvard University Press, 1966.

Copernicus, Nicolaus. *On the Revolutions*. In *Complete Works*. Edited by Jerzy Dobrzycki, translated by Edward Rosen. Vol. 2. Warsaw: Polish Scientific Publishers, 1978.

Descartes, René. *Traité de l'homme*. In *Oeuvres de Descartes*. Edited by Charles Adam and Paul Tannery. Vol. 11. Facsimile reprint. Paris: Vrin, 1967.

de Fontenelle, Bernard le Bovier. *Conversations on the Plurality of Worlds*. Edited by Nina Rattner Gelbart, translated by H. A. Hargreaves. Berkeley: University of California Press, 1990.

——. *Entretiens sur la pluralité des mondes*. Edited by Alexandre Calame. Paris: Librairie Marcel Didier, 1966.

Galilei, Galileo. *The Assayer*. In *The Controversy on the Comets of 1618*. Translated by Stillman Drake and C. D. O'Mailley. Philadelphia: University of Pennsylvania Press, 1960, 183-84.

——. *Dialogue Concerning the Two Chief World Systems*. Translated by Stillman Drake (Berkeley: University of California Press, 1967).

——. *Le méssager céleste*. Edited by Isabelle Pantin. Paris: Belles Lettres, 1992.

——. *Opere*. Edited by Antonio Favaro. 20 vols. Firenze: Barbèra, 1890-1909.

——. *Sidereus nuncius*. Translated by Albert Van Helden. Chicago: University of Chicago Press, 1989.

——. *Sidereus nuncius*. Edited by Andrea Battistini, translated by Maria Timpanaro Cardini. Venice: Marsilio, 1993.

Gentleman, Francis. *A Trip to the Moon. Containing an Account of the Island of NOIBLA. Its Inhabitants, Religious and Political Customs, etc.* (1764). Facsimile reprint. New York and London: Garland, 1974.

Hobbes, Thomas. *Leviathan, ore The Matter, Forme & Power of a Common-Wealth Ecclesiasticall and Civill*. London, 1651. First edition.

Horace. *Satire, Epistles, Ars poetica*. Translated by H. Rushton Fairclough. Loeb Classical Library. Cambridge, MA: Harvard University Press, 1961.

Huygens, Christiaan. *The Celestial Worlds Discovered* [*Cosmotheoros*]. (1698). Facsimile reprint. London: Frank Cass & Co., 1968.

——. *Christiani Hugenii Kosmotheoros, sive De Terris Coelestibus, earumque ornatu, conjecturae*. In *Oeuvres complètes de Christiaan Huygens*. Published by the Société Hollandaise des Sciences. The Hague: Martinus Nijhoff, 1944.

Kepler, Johannes. *Discussione col Nuncio Sidereo*. Edited by Elio Pasoli and Giorgio Tabarroni. Torino: Bottega d'Erasmo, 1972.

——. *Gesammelte Werke*. Edited by W. von Dyck and M. Caspar *et al.* 18 vols. Munich: Beck, 1937-.

——. *Kepler's Conversation with Galileo's Sidereal Messenger*. Edited and translated by Edward Rosen. New York: Johnson Reprint Corporation, 1965.

——. *Kepler's Dream*. Edited by John Lear, translated by Patricia Frueh Kirkwood. Berkeley and Los Angeles: University of California Press, 1965.

——. *Kepler's Somnium or Posthumous Work on Lunar Astronomy*. Edited and translated by Edward Rosen. Madison: University of Wisconsin Press, 1967.

——. *Keplers Traum vom Mond*. Edited and translated by Ludwig Günther. Leipzig, 1898.

Leibniz, Gottfried Wilhelm. *Essais de Théodicée, sur la bonté de dieu, la liberté de l'homme, et l'origine du mal.* Edited by J. Brunschwig. Paris, 1969.
Lucian, *True History.* In *Satirical Sketches.* Translated by Paul Turner. Bloomington: Indiana University Press, 1990.
Melville, Herman. "Bartleby, the Scrivener." *The Writings of Herman Melville.* Vol. 9. Evanston and Chicago: Northwestern University Press and The Newberry Library, 1987.
More, Henry. *Philosophical Poems of Henry More.* Edited by Geoffrey Bullough. Manchester: Manchester University Press, 1931.
de Patot, Tyssot. *Voyages et avantures de Jacques Massé.* Cologne, 1710.
Plato, *Timaeus.* In *Plato in Twelve Volumes.* Edited and translated by the Rev. R. G. Bury. Vol. 9. Loeb Classical Library. Cambridge, MA: Harvard University Press, 1965.
Plutarch. *Concerning the Face which Appears in the Orb of the Moon.* In *Moralia.* Edited and translated by Harold Cherniss, Vol. 12. Loeb Classical Library. Cambridge, MA: Harvard University Press, 1957.
Russen, David. *Iter lunare: Or a Voyage to the Moon.* London: 1707.
Schelling, F. W. J. *System des transzendentalen Idealismus.* Edited by Horst D. Brandt and Peter Müller. Hamburg: Meiner, 1992.
Tesauro, Emanuele. *Il cannocchiale aristotelico.* Edited by August Buck. Facsimile reprint. Bad Homburg: Gehlen, 1968.
Veiras, Denis. *Histoire des Sévarambes.* (1677-1679). Introduced by Raymond Trousson. Facsimile reprint. Geneva: Slatkine Reprints, 1979.
Wilkins, John. *The Discovery of a World in the Moone.* (1638). Facsimile reprint. Hildesheim and New York: Georg Olms Verlag, 1981.

Secondary Sources:

Arbib, Michael A. and Mary B. Hesse. *The Construction of Reality.* Cambridge: Cambridge University Press, 1986.
Ariew, Roger. "Galileo's Lunar Observations in the Context of Medieval Lunar Theory." *Studies in the History and Philosophy of Science* 15, no. 3 (1984): 213-26.
Ashworth Jr., William B. "Light of Reason, Light of Nature: Catholic and Protestant Metaphors of Scientific Knowledge." *Science in Context* 3, no. 1 (1989): 89-107.
Bagrow, Leo. *History of Cartography.* 2nd edition. Revised by R.A. Slelton. Chicago: Precedent Publishing, 1985.
Beer, Gilian. "Problems of Description in the Language of Discovery." *One Culture: Essays in Science and Literature.* Edited by George Levine. Madison: University of Wisconsin Press, 1987. 35-58.
Benedetti, Carla. *Pasolini contro Calvino. Per una letteratura impura.* Torino: Bollati Boringhieri, 1998.
Biagioli, Mario. *Galileo Courtier: The Practice of Science in The Culture of Absolutism.* Chicago: University of Chicago Press, 1993.
——. "Galileo's System of Patronage." *History of Science* 28 (1990): 1-62.
——. "Galileo the Emblem Maker." *Isis* 81 (1990): 230-58.

Biasin, Gian-Paolo. "Italo Calvino in Mexico: Food and Lovers, Tourists and Cannibals." *PMLA* 108, no. 1 (1993): 72-88.

Black, Max. "Metaphor." *Models and Metaphors*. Ithaca and London: Cornell University Press, 1962.

——. "More about Metaphor." *Metaphor and Thought*. Edited by Andrew Ortony. Cambridge: Cambridge University Press, 1979. 19-43.

Bloom, Terrie F. "Borrowed Perceptions: Harriot's Maps of the Moon." *Journal for the History of Astronomy* 9 (1978): 117-22.

Blumenberg, Hans. *Paradigmen zu einer Metaphorologie*. Bonn: Bouvier, 1960.

Bono, James. "Science, Discourse, and Literature: the Role/Rule of Metaphor in Science." *Literature and Science: Theory and Practice*. Edited by Stuart Peterfreund. Boston: Northwestern University Press, 1990. 59-89.

——. *The Word of God and the Languages of Men: Interpreting Nature in Modern Science and Medicine*. Vol. 1: *Ficino to Descartes*. Madison: University of Wisconsin Press, 1995.

Boyd, Richard. "Metaphor and Theory Change." *Metaphor and Thought*. Edited by Andrew Ortony. Cambridge: Cambridge University Press, 1979. 356-408.

Brodsky Lacour, Claudia. *Lines of Thought: Discourse, Architectonics, and the Origin of Modern Philosophy*. Durham and London: Duke University Press, 1996.

Brown, Harold I. "Galileo on the Telescope and the Eye." *Journal of the History of Ideas* 46 (1985): 487-501,

Calmé, Pierre Alain. *Un autre Descartes: le philosophe et son langage*. Paris: Vrin, 1980.

Carr, Thomas M. *Descartes and the Resilience of Rhetoric*. Carbondale and Edwardsville: Southern Illinois University Press, 1990.

Casini, Paolo. "Plutarco, Galileo e la faccia della luna." *Intersezioni* 4, no. 2 (1984): 397-404.

Cenerini, Lucia. "Metafora in utopia." *Il romanzo barocco tra Italia e Francia*. Rome: Bulzoni, 1980. 227-37.

Christianson, Gale E. "Kepler's *Somnium*: Science Fiction and the Renaissance Scientist." *Science-Fiction Studies* 4 (1977): 79-90.

Classen, J. "The First Map of the Moon," *Sky and Telescope* 37, no. 1 (1969): 82-84.

Cocking, J. M. *Imagination: A Study in the History of Ideas*. London and New York: Routledge, 1991.

Cohen, I. Bernard. "What Galileo Saw: The Experience of Looking through a Telescope." *From Galileo's 'Occhialino' to Optoelectronics*. International Conference at the University of Padova, 1992. 445-72.

Dagen, Jean. "Réflexions sur les mondes de Fontenelle." *Littératures classiques* 22 (1994): 127-44.

Dear, Peter. *Discipline & Experience: The Mathematical Way in the Scientific Revolution*. Chicago: University of Chicago Press, 1995.

——. "Jesuit Mathematical Science and the Reconstition of Experience in the Early Seventeenth Century." *Studies in History and Philosophy of Science* 18 (1987): 133-175.

——. *Revolutionizing the Sciences: European Knowledge and Its Ambitions, 1500-1700*. London: Pelgrave, 2001.

——. "*Totius in Verba*: Rhetoric and Authority in the Early Royal Society." *Isis* 76 (1985): 145-161.

de Man, Paul. "The Epistemology of Metaphor." *On Metaphor*. Edited by Sheldon Sacks. Chicago: University of Chicago Press, 1979.

Dick, Stephen J. *Plurality of Worlds: The Origins of the Extraterrestrial Life Debate from Democritus to Kant*. Cambridge: Cambridge University Press, 1982.

Drake, Stillman. "Galileo's Language: Mathematics and Poetry in a New Science." *Yale French Studies* 49 (1973): 13-27

——. "The Starry Messenger." *Isis* 49, no. 3 (1958): 346-47.

——, ed. *Discoveries and Opinions of Galileo*. Garden City, NJ: Doubleday, 1957.

Eamon, William. "Court, Academy, and Printing House: Patronage and Scientific Careers in Late Renaissance Italy." *Patronage and Institutions: Science, Technology and Medicine at the European Court, 1500-1750*. Edited by Bruce T. Moran. Woodbridge, Suffolk: The Boydell Press, 1991. 25-50.

Edgerton, Samuel J. "Galileo, Florentine 'Disegno', and the 'Strange Spottednesse' of the Moon." *Art Bulletin* 44 (1984): 225-32.

——. *The Heritage of Giotto's Geometry*. Ithaca and London: Cornell University Press, 1991.

Ferris, Timothy. *Coming of Age in the Milky Way*. New York: Doubleday, 1988.

Feyerabend, Paul. *Against Method*. 3rd edition. London and New York: Verso, 1994.

——. *Farewell to Reason*. London and New York: Verso, 1987.

Finocchiaro, Maurice. *The Galileo Affair*. Berkeley: University of California Press, 1989.

——. *Galileo and the Art of Reasoning: Rhetorical Foundations of Logic and Scientific Method*. Dordrecht and Boston: Reidel, 1980.

Foucault, Michel. "Of Other Spaces." Translated by Jay Miskowiec. *Diacritics* 16, no. 1 (1986): 22-27.

——. "The Order of Discourse." *Untying the Text: A Post-structuralist Reader*. Edited by Robert Young. London: Routledge and Kegan Paul, 48-78.

——. *The Order of Things: An Archaeology of the Human Sciences*. New York: Vintage Books, 1994.

Frey, Hans-Jost. *Die Autorität der Sprache*. Lana, Vienna and Zurich: Edition Howeg, 1999.

Funkenstein, Amos. *Theology and the Scientific Imagination from the Late Middle Ages to the XVIIth Century*. Princeton: Princeton University Press, 1986.

Gingerich, Owen. "Dissertatio cum Professore Righini et Sidereo Nuncio." *Reason, Experiment, and Mysticism in the Scientific Revolution*. Edited by Maria Luisa Righini Bonelli and William R. Shea. New York: Science History Publications, 1975. 77-88.

Gingerich, Owen and Ewan A. Whitaker. "Galileo's Lunar Observations and the Dating of the Composition of the *Sidereus nuncius*." *Journal of the History of Astronomy* 9 (1978): 155-69.

Gross, Alan. *The Rhetoric of Science*. Cambridge: Harvard University Press, 1990.

Güntert, Georges. *L'Orlando furioso*. Script of lecture. University of Zürich, summer 1993. Unpublished.

Guthke, Karl. *Der Mythos der Neuzeit: Das Thema der Mehrheit der Welten in der Literatur- und Geistesgeschichte von der kopernikanischen Wende bis zur Science Fiction*. Berne and Munich: Franke, 1983.

Hallyn, Fernand. *The Poetic Structure of the World: Copernicus and Kepler*. New York: Zone Books, 1990.

——. "Le Songe de Kepler." *Bibliothèque d'humanisme et renaissance, travaux et documents* 42 (1980): 329-47.

Hesse, Mary B. "The Cognitive Claims of Metaphor." *Metaphor and Religion*. Edited by J. P. van Noppen. Brussels: Free University of Brussels Press, 1984.

——. *Models and Analogies in Science*. Notre Dame: Notre Dame University Press, 1966.

Hetherington, Norriss S. *Science and Objectivity: Episodes in the History of Astronomy*. Ames: Iowa State University Press, 1988.

Hoch, Christoph. "Eudossias Teppich: Zu Theorie und Praxis intertextuellen Erzählens in Italo Calvinos *Le città invisibili*." *Romanische Forschungen* 106 (1994): 187-210.

Hoffman, Robert R. "Some Implications of Metaphor for Philosophy and Psychology of Science." *The Ubiquity of Metaphor: Metaphor in Language and Thought*. Current Issues in Linguistic Theory 29. Edited by Wolf Paprotté and René Dirven. Amsterdam and Philadelphia: John Benjamins, 1985. 327-80.

Howells, Robin. "The Principle of Mobility in Fontenelle's *Entretiens sur la pluralité des mondes*." *French Studies* 46, no 2 (1992): 129-43.

James, Carol P. "The Fragmentation of Allegory in Calvino's *Invisible Cities*." *The Review of Contemporary Fiction* 6, no. 2 (1986): 88-94.

James, E. D. "Fontenelle's Entretiens sur la pluralité des mondes and their Intellectual Context." *Actes de Columbus: Racine. Fontenelle, Entretiens sur la pluralité des monde:. Histoire et littérature*. Edited by Charles G. S. Williams. Paris: Papers on French Seventeenth Century Literature, 1990. 133-47.

Johnson, Mark, ed. *Philosophical Perspectives on Metaphor*. Minneapolis: University of Minnesota Press, 1981.

Jordanova, Ludmilla, ed. *Languages of Nature: Critical Essays on Science and Literature*. New Brunswick, NJ: Rutgers University Press, 1986.

——. *Sexual Visions: Images of Gender in Science and Medicine between the Eighteenth and Twentieth Centuries*. New York: Harvester Wheatsheaf, 1989.

Kearney, Richard. *The Wake of Imagination: Ideas of Creativity in Western Culture*. London: Hutchinson, 1988.

Koestler, Arthur. *The Sleepwalkers: A History of Man's Changing Vision of the Universe*. London: Penguin, 1989.

Kopal, Zdenek. *The Moon*. Dordrecht: Reidel, 1969.

Krappe, Alexandre H. *La genèse des mythes*. Paris: Payot, 1938.

Kuhn, Thomas. *The Copernican Revolution: Planetary Astronomy in the Development of Western Thought*. Cambridge: Harvard University Press, 1957.

——. "Metaphor in Science." *Metaphor and Thought*. Cambridge: Cambridge University Press, 1979. 409-19.

——. *The Structure of Scientific Revolutions*. Chicago: Chicago University Press, 1962.

Lakoff, George and Mark Johnson. *Metaphors We Live By*. Chicago: University of Chicago Press, 1980.

de Lauretis, Teresa. *Alice Doesn't: Feminism, Semiotics and Cinema*. Bloomington: Indiana University Press, 1984.

Leatherdale, W. M. *The Role of Analogy: Model and Metaphor in Science*. Amsterdam: North-Holland Press, 1974.

Lessle, Christine. *Weltreflexion und Weltlektüre in Italo Calvinos erzählerischem Spätwerk*. Bonn: Romanischer Verlag, 1992.

Lindberg, David C. and Robert S. Westman, eds. *Reappraisals of the Scientific Revolution*. Cambridge: Cambridge University Press.

Lovejoy, Arthur O. *The Great Chain of Being: A Study of the History of an Idea*. The William James Lectures Delivered at Harvard University, 1933. Reprint, Cambridge: Harvard University Press, 1976.

MacCormac, Earl. *Metaphor and Myth in Science and Religion*. Durham: Duke University Press, 1976.

Menzel, Donald H. "Kepler's Place in Science Fiction." *Vistas in Astronomy* 18 (1975): 896-904.

Montgomery, Scott L. *The Scientific Voice*. New York and London: Guildford Press, 1996,

Moss, Jean Dietz. *Novelties in the Heavens: Rhetoric and Science in the Copernican Controversy*. Chicago: University of Chicago Press, 1993.

Musacchio, Enrico. *Amore, ragione e follia: una rilettura dell'*Orlando furioso. Rome: Bulzoni, 1983.

Nicolson, Marjorie Hope. *Science and Imagination*. Ithaca and London: Cornell University Press, 1962.

——. *Voyages to the Moon*. New York: Macmillan, 1948.

——. "A World in the Moon: A Study of the Changing Attitude toward the Moon in the Seventeenth and Eighteenth Centuries." *Smith College Studies in Modern Languages* 17, no. 2 (1936): 1-72.

Olson, Richard, ed. *Science as Metaphor: The Historical Role of Scientific Theories in Forming Western Culture*. Belmont, CA: Wadsworth, 1971.

Ostrow, Steven F. *Art and Spirituality in Counter-Reformation Rome: The Sistine and Pauline Chapels in S. Maria Maggiore*. Cambridge: Cambridge University Press, 1996.

——. "Cigoli's *Immacolata* and Galileo's Moon: Astronomy and the Virgin in Early Seicento Rome." *Art Bulletin* 77, no. 2 (1996): 218-35.

Piazza, Gianguido. "La Luna della Luna. Metafore, congetture e scoperte della moderna astronomia planetaria." *La Luna allo specchio*. Edited by Nadia Minerva. Bologna: Pàtron, 1990. 13-34.

Raimondi, Ezio. "La nuova scienza e la visione degli oggetti." *Lettere italiane* 21, no. 3 (1969): 265-305.

Redondi, Pietro. *Galileo Heretic*. Translated by Raymond Rosenthal. Princeton, NJ: Princeton University Press, 1987.

Reeves, Eileen. *Painting the Heavens: Art and Science in the Age of Galileo*. Princeton, NJ: Princeton University Press, 1997.

——. "The Rhetoric of Optics: Perspectives on Galileo and Tesauro." *Stanford Italian Review* 7 (1987): 129-45.

Reiss, Timothy J. *Against Autonomy: Global Dialectics of Cultural Exchange*. Stanford: Stanford University Press, forthcoming 2002.

——. *The Discourse of Modernism*. Ithaca and London: Cornell University Press, 1982.

——. "Espaces de la pensée discursive: le cas Galilée et la science classique." *Revue de Synthèse* 85-86 (Jan. – June 1977): 5-47.

——. *Knowledge, Discovery and Imagination in Early Modern Europe: The Rise of Aesthetic Rationalism*. Cambridge: Cambridge University Press, 1997.

——. *The Meaning of Literature*. Ithaca and London: Cornell University Press, 1992.

Richards, I. A. *The Philosophy of Rhetoric*. Oxford: Oxford University Press, 1936.

Ricoeur, Paul. "The Metaphorical Process as Cognition, Imagination, and Feeling." *Philosophical Perspectives on Metaphor*. Edited by Mark Johnson. Minneapolis: University of Minnesota Press, 1981.

——. *The Rule of Metaphor*. Translated by Robert Czerny. Toronto: University of Toronto Press, 1977.

Righini, Guglielmo. "New Light on Galileo's Lunar Observation." *Reason, Experiment, and Mysticism in the Scientific Revolution*. Edited by Maria Luisa Righini Bonelli and William R. Shea. New York: Science History Publications, 1975. 59-76.

Rosen, Edward. "The Title of Galileo's *Sidereus nuncius*." *Isis* 41 (1950): 287-89.

——. "Review: Stillman Drake's Discoveries and Opinions of Galileo." *Journal of the History of Ideas 18* (1957): 439-48.

Rossi, Paolo. *Philosophy, Technology, and the Arts in the Early Modern Era*. Translated by Salvator Attanasio. New York: Harper and Row, 1970.

Russo, François. "Note sur la traduction du titre de l'ouvrage de Galilée, *Sidereus nuncius*." *Revue d'histoire des sciences* 20 (1967): 67-69.

Sartre, Jean-Paul. *L'être et le néant; essai d'ontologie phénoménologique*. Paris: Gallimard, 1957.

Schaffer, Simon. "On Astronomical Drawing." *Picturing Science, Producing Art*. Edited by Caroline A. Jones and Peter Galison. New York and London: Routledge, 1998. 441-74.

——. "Self-Evidence." *Critical Inquiry* 19 (1992): 327-62.

Schnyder, Rudolf, ed. *Zürcher Staatsaltertümer. Der Zürcher Staat im 17. Jahrhundert*. Berne: Paul Haupt, 1975.

Shapin, Steven. "Pump and Circumstance: Robert Boyle's Literary Technology." *Social Studies of Science* 14 (1984): 481-520.

——. "Robert Boyle and Mathematics: Reality, Representation, and Experimental Practice," *Science in Context* 2, no. 1 (1988): 23-58.

——. *The Scientific Revolution*. Chicago: University of Chicago Press, 1996.

Shapin, Steven and Simon Schaffer. *Leviathan and the Air-Pump: Hobbes, Boyle, and the Experimental Life*. Princeton, NJ: Princeton University Press, 1985.

Shapiro, Barbara, J. *John Wilkins 1614-72: An Intellectual Biography*. Berkeley: University of California Press, 1969.

——. *Probability and Certainty in Seventeenth-Century England: A Study of the Relationships between Natrual Science, Religion, History, Law, and Literature.* Princeton, NJ: Princeton University Press, 1983.

Simon, Gérard. *Kepler, astronome, astrologue*. Paris: Gallimard, 1979.

Slawinski, Maurice. "Rhetoric and Science/Rhetoric of Science/Rhetoric as Science." *Science, Culture and Popular Belief in Renaissance Europe*. Edited by Stephen Pumfrey, Paolo Rossi and Maurice Slawinski. Manchester: Manchester University Press, 1991. 71-99.

Smith, A. J. *Metaphysical Wit*. Cambridge: Cambridge University Press, 1991.

Solomon, Robert C. *The Passions*. Indianapolis: University of Indiana Press, 1993.

Soskice, Janet Martin and Rom Harré, "Metaphor in Science." *From a Metaphorical Point of View: A Multidisciplinary Approach to the Cognitive Content of Metaphor*. Edited by Zdravko Radman. Berlin: de Gruyter, 1995. 289-307.

Stepan, Nancy Leys. *'The Hour of Eugenics': Race, Gender, and Nation in Latin America*. Ithaca and London: Cornell University Press, 1991.

——. "Race and Gender: The Role of Analogy in Science." *Isis* 77 (1986): 261-77.

Stooke, Philip. "Mappaemundi and the Mirror in the Moon." *Cartographia* 29, no. 2 (1992): 20-30.

Sutton, John. *Philosophy and Memory Traces: Descartes to Connectionism.* Cambridge: Cambridge University Press, 1998.

Van Helden, Albert. "Galileo and Scheiner on Sunspots: A Case Study in the Visual Language of Astronomy." *Proceedings of the American Philosophical Society* 140, no. 3 (1995): 358-96.

——. "The Invention of the Telescope." *Proceedings of the American Philosophical Society* 67, no. 4 (1977): 5-67.

Van Den Abbeele, Georges. "*Fabula est Mundus*: On the Plurality of Fontenelle's 'Worlds'." *Actes de Columbus: Racine. Fontenelle, Entretiens sur la pluralité des mondes: Histoire et littérature*. Edited by Charles G. S. Williams. Paris: Papers on French Seventeenth Century Literature, 1990. 165-83.

Vickers, Brian. "Epideictic Rhetoric in Galileo's *Dialogo*." *Annali dell'Istituto e Museo di Storia della Scienza di Firenze* 8, no. 2 (1983): 70-102.

Warnock, Mary. *Imagination*. London: Faber & Faber, 1976.

Westman, Robert. "The Astronomer's Role in the Sixteenth Century: A Preliminary Study." *History of Science* 18 (1980): 105-47.

——. "The Copernicans and the Churches." *God and Nature*. Edited by David C. Lindberg and Ronald L. Numbers. Berkeley: University of California Press, 1986. 76-113.

Whitaker, Ewen A. "Selenography in the Seventeenth Century." *The General History of Astronomy*. Edited by René Taton. Vol. 2. Cambridge: Cambridge University Press, 1989.

White, Alan R. *The Language of Imagination*. London: Blackwell, 1990.

Wiggins, Peter DeSa. *Figures in Ariosto's Tapestry: Character and Design in the* Orlando furioso. Baltimore and London: Johns Hopkins University Press, 1986.

Williams, Charles G. S., ed. *Actes de Columbus: Racine. Fontenelle, Entretiens sur la pluralité des mondes: Histoire et littérature*. Paris: Papers on Seventeenth-Century French Literature, 1990.

Winkler, Mary G. and Albert Van Helden. "Johannes Hevelius and the Visual Language of Astronomy." *Renaissance and Revolution: Humanists, Scholars, Craftsmen, and Natural Philosophers in Early Modern Europe*. Edited by J. V. Field and Frank A. J. L. James. Cambridge: Cambridge University Press, 1994. 97-116.

——. "Representing the Heavens: Galileo and Visual Astronomy." *Isis* 83 (1992): 195-217.

Yates, Frances. *The Art of Memory*. London: Routledge and Kegan, 1966.

——. *Giordano Bruno and the Hermetic Tradition*. Chicago: University of Chicago Press, 1964.

Zeilik, Michael and John Gaustad, eds. *Astronomy: The Cosmic Perspective*. New York: Harper & Row, 1983.

www.ingramcontent.com/pod-product-compliance
Ingram Content Group UK Ltd.
Pitfield, Milton Keynes, MK11 3LW, UK
UKHW050759210926
15297UKWH00013B/160

9 789042 015784